ANALYSIS OF NORMAL
and
ABNORMAL CELL GROWTH

ANALYSIS OF NORMAL
and
ABNORMAL CELL GROWTH

Model-System Formulations and Analog Computer Studies

FERDINAND HEINMETS
Biophysics Laboratory
Pioneering Research Division
U. S. Army Natick Laboratories
Natick, Massachusetts

Springer Science+Business Media, LLC
1966

ISBN 978-1-4899-6273-7   ISBN 978-1-4899-6594-3 (eBook)
DOI 10.1007/978-1-4899-6594-3

The findings in this publication are not to be construed as an official statement of the Army position.

Library of Congress Catalog Card Number 66-11882

© 1966 Springer Science+Business Media New York
Originally published by Plenum Press in 1966
Softcover reprint of the hardcover 1st edition 1966

This book is dedicated to my wife and to my
sons, Hilary and Julian, who often forbore
the burden of its preparation

Preface

This book represents an attempt to explore a complex biological problem. The subject is analyzed from a theoretical point of view, and various procedures are utilized to gain insight into mechanisms of the basic processes. Quantitative study of a model-system is carried out on an analog computer, which serves as a mathematical tool, but not as a simulating device. Since the book is meant primarily for all students and researchers in biology who are interested in cellular problems at a very basic level, details of computer technological procedures as well as mathematical formulations have been omitted. The main aspects of this book are organizational dynamics of cellular growth and cellular interactions.

This book can be read by any competent student of biology; no specific mathematical background is necessary for an adequate comprehension. Those who are not familiar with differential equations may ignore the ones presented here and follow the flow equations in the tables in the text. Also, no background is required in analog computer techniques for an understanding of the basic analysis. This work is projective in character, and the subject, in many areas, is treated in a highly speculative manner, especially when the phenomenon of abnormal growth and cancer is discussed. The primary purpose here is to stimulate new thinking and experimental approaches to this complex research problem. It may appear that the author has been too confident in deriving various conclusions.   This may have resulted from an overdose of self-encouragement during the work, where many difficulties were encountered in designing the model-system, as well as in solving the mathematical problems.   Therefore, this work carries a personal stamp and should not be taken too dogmatically.   It is hoped, however, that it might induce the reader to consider biological processes from a more quantitative point of view.

F. HEINMETS

January, 1966

Acknowledgments

The invaluable help of Mr. F. Melanson (GPS Instrument Co., Newton, Mass., U.S.A.) in applying the analog computer techniques for solving the problem is recognized, and his patient willingness to try something "new" again is highly appreciated. We are also thankful to Mr. P. A. Holst of Chr. Michelsen's Institute, Bergen, Norway, for part of the computer programming. The labor of preparing the unwieldy manuscript for publication was performed with skill and patience by Mary T. O'Brien and Carolyn Nalbandian. We are greatly indebted to J. L. Whelan and his staff, H. R. Perry, C. Kontos and A. J. Forcier, Audiovisual Aids Office, for preparing graphs for the printing.

I highly appreciate the moral and administrative support of Dr. S. D. Bailey and Dr. G. R. Mandels in carrying out this research problem. I am also greatly indebted to Dr. A. J. Rosenberg of Institut de Biologie Physico-Chimique, Paris, France, for discussions and encouragement to initiate this study. Finally, I would like to recognize the help and interest of Dr. D. H. K. Lee for making it possible for me to carry out the studies in the field of theoretical biology.

Contents

PART II Descriptive Analysis of an Advanced Cellular
Model-System

PART III Analysis of the Cancer Problem

Introduction

It appears that there has been more knowledge and understanding gained in the mechanisms of basic biological processes
during the last decade than during the whole preceding
century. Several reasons for this may be the accumulation of
enormous experimental data in the field of descriptive biology,
the development of basic scientific disciplines which could be
utilized for the interpretation of operational mechanisms of
various processes, and, finally, the broadening of the educa-
tional scope of scientific training, which permits the scientist
to search and observe in a more integrated manner. Since
biological processes in general are extremely complex in
terms of molecular processes, the integrative and quantitative
approach is essential for the further progress in biological
research.

There has been extremely rapid progress in the area of
molecular biology, which deals with basic functional entities of
the cell. Not only have the functional role and basic properties
of principal cellular elements, such as DNA, RNA proteins,
etc., been elucidated, but in many cases the exact chemical formula and structural characteristics have been determined. As a matter of fact, the progress in this all-important
research area has been and still is so rapid that it is difficult
for the individual scientist to follow contemporary progress
completely. It seems obvious that the field of molecular
biology will continue to progress rapidly, and the question can

1

be asked whether this basic information could be utilized to expand the scope of biological research. It seems obvious to us that the next step, after having gained a certain amount of information on the properties of basic structural entities of the cell, would be the study of the integrated interrelation of these entities as a total functional system. In this way, it would be possible to gain insight and elucidate the basic cellular processes, such as growth, multiplication, differentiation, and so forth. In order to carry out such work, it seems essential to establish model-systems containing basic functional entities, to formulate the system mathematically, and to carry out a quantitative analysis of an operational system. This would permit not only the study of basic integrated processes, but also such processes in conditions of stress when various external agents are introduced into the system.

Growth is a basic property of biological species, and growth coupled with the division process leads to the increase of population. Cellular growth can be considered from various points of view—for example, as it exists at the present moment, thus ignoring evolutionary aspects, since evolution represents the development of new biological systems and new biological properties. It is not known how biological systems have been initiated, or how future systems will be developed; both are subjects for extensive speculation. In the present work we shall consider the growth process only as it exists today and ignore the evolutionary aspects because these are so slow that they do not influence the growth process in the framework of the time under consideration.

The complexity of biological systems is obvious. What is not so obvious is how to gain more insight into detailed mechanisms inherent in cell growth and integrated processes. Collecting experimental data from isolated experiments yields important raw material for the synthesis of more complex systems. However, detailed information obtained from experiments which are performed in conditions which are not equivalent to the cellular environment represent only a limited type of information, and such information has to be integrated into a larger framework of cellular operations. Current analysis of cellular growth processes in the literature is essentially carried out by a descriptive process. However, verbal nonquantitative analysis of nonlinear systems is not adequate. The usual intellectual reasoning process involves only a few steps of linear logic. Biological processes are essentially nonlinear. Furthermore, they represent a multiple network of interactions and, hence, application of linear logic into such a framework is leading to an <u>illusion of understanding</u> rather than a real understanding of the basic process. It is quite certain that thinking through a multiple-step nonlinear system is not a method for solving such a problem. Biological processes when considered at the molecular level involve a large number of functional entities, and a descriptive analysis of such a system is limited although it may be occasionally of some limited value. In order to comprehend the basic cellular processes, it is essential that relationships between individual functional units should be analyzed as a function of time, since the growth process represents the change of absolute concentration of those entities. Therefore, it is essential that formalized relationships be established between the operational elements of the system and that the problem be solved in a quantitative manner. In order to do this it is imperative that a model-system be established. In cellular metabolic systems there are a number of similar types of elementary units which are organized into complex patterns. It is considered that an analysis of a model-system which contains the principal functional entities such as DNA, RNA, and enzymes may lead to an understanding of growth processes and their regulation. Quantitative analysis is imperative in the study of complex regulatory processes. There are varieties of interaction between basic functional entities in biological processes. Some elements have antagonistic action while other elements have mutually supportive action. There are feedback processes which are positive and negative in character, and in the framework of such interactions we may have periodic phenomena. It is obvious that such complex processes can be analyzed only by quantitative techniques. Furthermore, we can expect that only by quantitative evaluation of functional processes can the conditions of imbalance be established and their effect on growth be determined. Such considerations are very important in processes of what we call abnormal growth. It is only by an understanding of normal cellular processes that one can gain insight into phenomena such as malignancy, cell alterations during the aging of species, drug-action, and radiation. All agents which interfere with the normal metabolic processes and synthesis can be completely understood only when the metabolic systems are represented in terms of relationships and interactions. This is of course also a prerequisite for developing a rational therapy against various cellular abnormalities caused by intracellular disorganization per se or caused by extracellular agents acting on an intracellular system via penetration into that medium.

Since we have established that in order to analyze cell growth and related phenomena it is essential to establish a model-system, the question will then arise as to how to build a model-system and what kind of information is required for it. One could consider two possible modes: (a) an intuitive one, in which various abstract considerations form the basis of the building of a model system; (b) the use of all available information which is experimental in nature and the synthesis from these data of a model-system for the analysis (such information would consist of all the data which appear in the literature). Since abstract considerations are arbitrary, we take the second consideration as the basis for model building. While the literature provides the basic information, it still remains for the author to integrate and organize such information into a functional system. In this process, many speculations often have to be made, and a good deal of the organization of the work may represent the point of view of the individual who performs the task; these circumstances are inevitable in performing this type of work. Nevertheless, this type of work carries a stamp of the author's individual opinion concerning the organization of cellular processes and may not necessarily meet with universal concurrence or appreciation.

The development and construction of a model-system is by no means an easy process. It would be desirable to have a very complete model which would help interpret many complex cellular processes. However, quantitative formulation is limited by many prohibitive factors. Consequently, the complexity of the model is limited to that level where the mathematical and computer techniques permit it to be carried out. In order to analyze various phenomena related to cell growth, we have decided to establish two models. The first model (I), which can be analyzed mathematically, is established only for cell growth, but not for cellular division. This system contains all basic functional entities which are operational in the cell, and they are organized in a framework of mutual interaction as known from the data in the literature. The second model (II) is a descriptive type. It permits a speculative analysis of cell growth and cell division. It includes trigger mechanisms for chromosomal division and cell division. It also permits a descriptive analysis in a more detailed manner on interaction of the intracellular system with interfering external agents which penetrate into the cell. We shall first develop and carry out calculations on the model-system for quantitative analysis; second, the descriptive model will be analyzed and discussed.

PART I

Quantitative Analysis of the Model-System

1. INTRODUCTION

The principal limiting factors in setting up a model-system are the following:

a. The capacity of the computer, i.e., whether it has a sufficient number of operational units and elements. Since the computer is used as a mathematical tool, but not as a simulating device, the number of differential equations which can be programmed on the computer is one of the limiting factors.

b. The ability of the operator to solve the problem. This may be considered one of the most important limiting factors. After a certain amount of experimentation, it was decided to establish a model-system which contained four basic functional genes which represent group-properties. This approach is justified if one considers that the group-property is based on the fact that all basic functional entities are built up with a certain number of integral building blocks, and a complete number of blocks have to be there in order to build a basic functional entity. If one building block is missing, a functional unit cannot be built. Consequently, the group-property will give a cross-behavior of the system. In the initial condition it is assumed that all functional entities have a relative value of unity and as a function of time all entities will start to grow. They reach approximately double value at the end of the generation time, which has been selected as a convenient time interval in the framework of computer observation time. It will be noted that all functional entities do not start to grow at the same rate. Instead, they each have different kinetic characteristics, and many entities may initially decline in absolute value and later grow more rapidly than other elements. The growth of the model-system is initiated from initial conditions by activating the external pool. This process is equivalent to taking cells and placing them in a nutrient medium and measuring their growth.

In general, the basic elements of the system are: genes, messengers (RNA), templates, enzymes, repressors, and activators. The model-system is so organized that a number of cellular processes and events can take place in conditions so that the coupling between these processes can have various degrees of freedom. There can be coupling processes which are strong and definitive or there may be couplings which are only inductive and need not always lead into subsequent coupled events. These processes will be described in detail later.

2. SPECIFIC FUNCTIONAL PHENOMENA

In the descriptive model, the separation of extracellular medium and intracellular organization is accomplished by a cell membrane which has certain transport characteristics. Intracellular growth processes are subject to control by the nuclear membrane; however, there is little information on the specific characteristics of transport properties, and the problem can be dealt with only in a general manner. It is considered that the external pool P_e is a constant during the experiment, but its value can be varied and the effect of pool concentration on growth can be studied.

There is evidence that RNA synthesis is probably under the control of amino acids. The subject has been discussed extensively by Maale and Kurland [3]. It is evident that the mechanism of such control is far from clear and there may be multiple sets of steps involved in such a process. At the current level of model-system formulation, it is impossible to introduce such a complex type of regulatory mechanism, especially where details of the basic process are unknown and

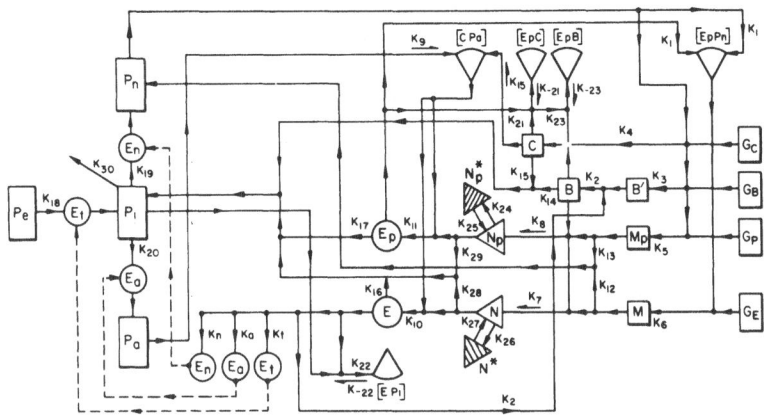

Fig. 1. A model-system for cell growth analysis.

open for speculation. In order to reduce an intricate set of reaction patterns into a few sets of equations, the system has to be thoroughly understood. This data reduction is essential, since the model-system analysis on the computer is confined to a simple organizational pattern which contains only the basic functional entities required for the operational system. Therefore, it is essential that more experimental information be available for a more advanced model-system formulation. The present model (Fig. 1) has also some regulatory features involving amino acids, but these are more limited. Amino acid concentration, which is identical to free pool P_a in our model system, has an indirect regulatory function. Pool P_a interacts with transport RNA C and forms the complex $[CP_a]$ which is utilized for enzyme synthesis. The complex $[CP_a]$ formation is irreversible in our system, and this feature guarantees that free P_a is rapidly used up. As we see in the rapidly growing cell, pool P_a is always present in the system, and the problem of growth initiation is not important while the cell is already rapidly growing. However, it is important in cellular dormancy.

3. FORMULATION AND DESCRIPTION OF MODEL-SYSTEM I

The terminology used in model building is represented in Table I and the flow schemes in Table II. A schematic diagram

TABLE I. Terminology and Symbols

Pools

P_e = Extracellular nutrient pool

P_i = General intracellular metabolic pool

P_a = Amino acid pool for protein synthesis

P_n = Nucleotide pool for RNA synthesis

Enzymes

E = Total protein

E_n = Enzymes which convert internal pool (P_i) into RNA precursors

E_a = Enzymes which convert internal pool (P_i) into amino acids

E_p = RNA polymerase for messenger RNA (M) synthesis

E_t = Enzymes which convert external pool (P_e) into internal pool (P_i)

Rate constant k_n, k_a, and k_t determine what fraction of total protein represents respective enzymes.

Genes

G_E = genes for messenger RNA (M) synthesis.

G_P = gene for messenger RNA (M_p) synthesis

G_B = genes for the synthesis of RNA fraction of ribosome

G_C = genes for transport RNA (C) synthesis

Messengers

M = Messenger (RNA) for protein (E) synthesis

M_p = Messenger (RNA) for E_p synthesis

B' = RNA fraction of ribosome

B = Ribosome

C = Transport RNA

N = Ribosome and messenger complex for protein (E) synthesis (template)

N_p = Ribosome and messenger complex for E_p synthesis (template)

N' = Inactive state of N

N'_p = Inactive state of N_p

s_i = Metabolite which converts templates N and N_p into inactive state

s'_i = Metabolite which converts inactive template N' and N'_p into active state

$k_1 \ldots k_n$ = Various rate constants

TABLE II. General Scheme

1. $E_p + P_n \xrightarrow{k_1} [E_pP_n]$

2. $B' + E \xrightarrow{k_2} B$

3. $G_B + P_n \xrightarrow{k_3} G_B + B'$

4. $G_C + P_n \xrightarrow{k_4} G_C + C$

5. $G_P + P_n \xrightarrow{k_5} G_P + M_p$

6. $G_E + [E_pP_n] \xrightarrow{k_6} G_E + E_p + M$

7. $B + M \xrightarrow{k_7} N$

8. $B + M_p \xrightarrow{k_8} N_p$

9. $C + P_a \xrightarrow{k_9} [CP_a]$

10. $N + [CP_a] \xrightarrow{k_{10}} B + C + M + E$

11. $N_p + [CP_a] \xrightarrow{k_{11}} B + M_p + C + E_p$

12. $M \xrightarrow{k_{12}} P_n$

13. $M_p \xrightarrow{k_{13}} P_n$

14. $B \xrightarrow{k_{14}} P_i$

15. $C \xrightarrow{k_{15}} P_i$

16. $E \xrightarrow{k_{16}} P_i$

17. $E_p \xrightarrow{k_{17}} P_i$

18. $P_e + E_t \xrightarrow{k'_{18}} P_i + E_t$

19. $P_i + E_n \xrightarrow{k'_{19}} P_n + E_n$

20. $P_i + E_a \xrightarrow{k'_{20}} P_a + E_a$

21. $E_p + C \underset{k_{-21}}{\overset{k_{21}}{\rightleftharpoons}} [E_pC]$

22. $E + P_i \underset{k_{-22}}{\overset{k_{22}}{\rightleftharpoons}} [EP_i]$

23. $E_p + B \underset{k_{-23}}{\overset{k_{23}}{\rightleftharpoons}} [E_pB]$

24. $N_p + s_i \xrightarrow{k_{24}} N'p$

25. $N'p + s'_i \xrightarrow{k_{25}} N_p$

26. $N + s_i \xrightarrow{k_{26}} N'$

27. $N' + s'_i \xrightarrow{k_{27}} N$

28. $N \xrightarrow{k_{28}} P_i$

29. $N_p \xrightarrow{k_{29}} P_i$

30. $P_i \xrightarrow{k_{30}} X$

where:

$E_n = k_n E \qquad k_{18} = k'_{18}\, k_t$

$E_t = k_t E \qquad k_{19} = k'_{19}\, k_n$

$E_a = k_a E \qquad k_{20} = k'_{20}\, k_a$

of the model is presented in Fig. 1. G_E represents a group of genes which are required for protein synthesis. Gene G_P is for the synthesis of the enzyme E_p, which is the polymerase for messenger M. Gene G_B produces RNA fraction of the ribosome, while the protein fraction of ribosome is provided by total protein E. Gene G_C produces the transport RNA. Let us review briefly the operational characteristics of a model-system. The first functional process can be considered to be the complexing of messenger M with ribosome B yielding the template N. Messenger M is produced by interaction of complex $[E_pP_n]$ with gene G_E. Enzyme synthesis results when templates interact with transport RNA and the amino acid pool complex $[CP_a]$. The total protein E contains three fractions: E_n, enzymes which convert internal pool P_i into RNA precursors; E_a, enzymes which convert internal pool P_i into amino acids; E_t, enzymes which convert external pool P_e into internal pool P_i. Rate constants (k_n, k_a, and k_t) determine what fraction of the total protein is converted to respective enzymes. It should be noted that gene G_E is the only gene where polymerase is operating in the messenger synthesis. Originally it was intended to have the same polymerase E_p to be operational with all genes. However, competition among various genes for the same polymerase made it difficult to organize a stable functional system on the computer. It seems that these genes should each have different kinds of polymerases. However, inclusion of four polymerases would have made the system extremely complex, and consequently other genes are considered to interact with pool P_n directly. Nevertheless, participation of a polymerase in forming messenger M for total protein will give system characteristics where a polymerase is part of the gene product. The formation of polymerase E_p follows generally the same line represented by total protein E synthesis. External nutrient pool P_e is converted into an internal pool P_i. In addition, part of the pool P_i leaks out from the cell via k_{30}, but the product X will not be associated with the external pool. The principal regulatory element in the system is the complexing of total protein with the internal pool (equation 22,

Table II). This complexing is reversible, and the ratio between the two rate constants, k_{22} and k_{-22}, will determine the degree of regulation at that level. Additional regulatory features are the complexing of polymerase with transport RNA (equation 21) and the complexing of polymerase with ribosomes (equation 23). All functional entities have different degrees of stability, and they decay into their respective pools. The system is growing, because there is an input via transport enzyme E_t from the external pool. In order to analyze cellular dormancy two additional entities are introduced into the system. It is visualized that template N can be converted into inactive form N' and template N_p into N_p'. This conversion can result from the action of metabolites or hormones and is considered to be reversible by activation process.

In order to analyze the type of model-system in a quantitative manner, it is essential that a systematic formulation be made. Consequently the flow scheme in Table II is organized into a set of simultaneous differential equations. These are represented in Table III. Subsequently, the differential equations are programmed for the analog computer (Fig. 2). This procedure is not further discussed here, since it has been analyzed in previous publications [1,2,]. When the program has been assembled on the analog computer, it is essential to establish a functional system. The principal characteristics of a functional system for the cell growth can be considered to be the requirement that during the generation time all cell components have to double their initial value. This is difficult to achieve, since there are a considerable number of interactions among various functional entities. As a consequence, by changing any individual rate constant, other functional entities change by different degrees. Furthermore, a change of some entity in one direction can produce changes in other entities to the opposite direction. The final condition that all entities will have doubled at the end of the generation time, while in a certain phase of generation time they may even decrease, is not easy to obtain. It requires a great deal of

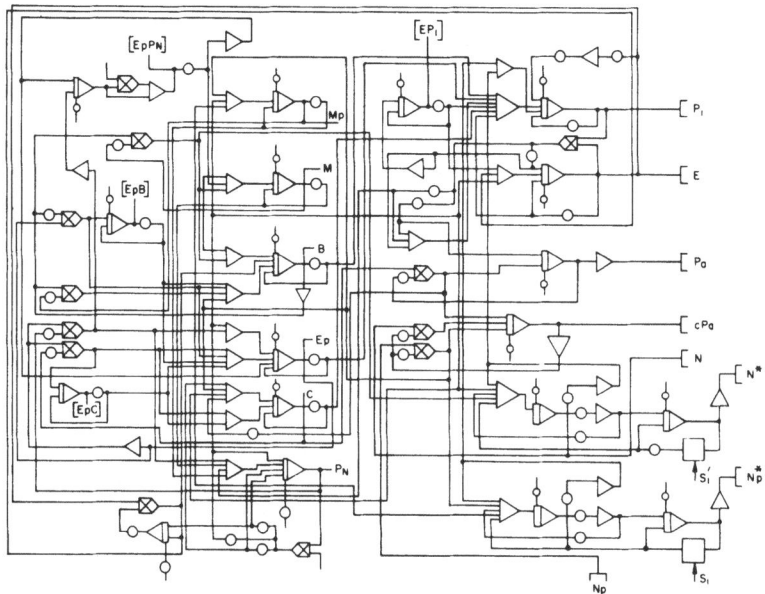

Fig. 2. Analog computer programming for differential equations (Table III).

both patience and experience in computer technology to organize a functional system of a model which contains many entities (31 rate constants). Only after tedious and prolonged trials is it possible to develop a functional system. The principal cause of difficulty lies in the fact that the system is highly nonlinear. As a matter of fact, it is a network of nonlinearities. This is of course the principal feature of the biological system, where many degrees of interdependence exist among functional entities. Consequently, quantitative study of the biological systems is extremely difficult, and new computer techniques must be developed. No attempt is made here to go into the details of analog techniques and the procedures used to obtain functional systems. This may be of interest only to specialists and not to biologists who are interested more in biological phenomena than in the organization of a computer model.

Once the functional system has been established on the computer, then it is possible to carry out experiments on the

TABLE III. Differential Equations

1. $\dot{E}_p = k_{11}N_p\,[C\,Pa] - k_{17}E_p - k_{21}E_p\,C + k_{-21}\,[E_pC] - k_{23}E_pB + k_{-23}[E_pB] - k_1E_pP_n + k_6G_E[E_pP_n]$

2. $\dot{P}_n = k_{19}E\,P_i - k_1E_pP_n + k_{12}M + k_{13}M_p - k_4G_CP_n - k_5\,G_PP_n - k_3G_BP_n$

3. $[\overline{\dot{E}_pP_n}] = k_1E_pP_n - k_6G_E[E_pP_n]$

4. $\dot{B}' = k_3G_BP_n - k_2B'\,E$

5. $\dot{B} = k_2B'\,E - k_7BM - k_8BM_p - k_{14}B + k_{10}N\,[C\,P_a] + k_{11}N_p\,[C\,Pa] - k_{23}\,E_pB + k_{-23}[\,E_pB]$

6. $\dot{C} = k_4G_CP_n - k_9CP_a + k_{10}N\,[C\,P_a] + k_{11}N_p[CP_a] - k_{15}C - k_{21}E_pC + k_{-21}\,[\,E_pC]$

7. $\dot{M} = k_6G_E\,[\,E_pP_n] - k_7BM - k_{12}M + k_{10}N\,[C\,P_a]$

8. $\dot{M}_p = k_5G_PP_n - k_8BM_p + k_{11}N_p[CP_a] - k_{13}\,M_p$

9. $\dot{N} = k_7BM - k_{26}s_iN + k_{27}\,s'_i\,N' - k_{10}N\,[CP_a] - k_{28}N$

10. $\dot{N}_p = k_8B\,M_p - k_{24}s_i\,N_p + k_{25}s'_i\,N'_p - k_{11}N_p\,[C\,P_a] - k_{29}N_p$

11. $\dot{N}' = k_{26}\,s_iN - k_{27}s'_iN'$

12. $\dot{N}'_p = k_{24}s_i\,N_p - k_{25}s'_i\,N'_p$

13. $\dot{P}_a = k_{20}P_iE - k_9C\,P_a$

14. $[\overline{\dot{C\,P_a}}] = k_9C\,P_a - k_{10}N\,[C\,P_a] - k_{11}N_p\,[C\,P_a]$

15. $\dot{E} = k_{10}N\,[CP_a] - k_{16}E - k_{22}E\,P_i + k_{-22}[EP_i] - k_2EB'$

16. $\dot{P}_i = k_{18}E\,P_e - k_{19}E\,P_i - k_{20}E\,P_i - k_{22}E\,P_i + k_{-22}[\,E\,P_i] + k_{14}B + k_{15}C + k_{16}E + k_{17}E_p + k_{28}N + k_{29}N_p - k_{30}P_i$

17. $[\overline{\dot{E_pC}}] = k_{21}E_pC - k_{-21}[E_pC]$

18. $[\overline{\dot{E_pB}}] = k_{23}E_pB - k_{-23}[E_pB]$

19. $[\overline{\dot{E\,P_i}}] = k_{22}E\,P_i - k_{-22}[\,E\,P_i]$

model-system and observe kinetic behavior of the system on an oscilloscope screen. It is instructive to observe the results when certain rate constants of the system are changed or the initial conditions altered. Varieties of experiments of this nature can be carried out and subsequently recorded for permanent records. Laboratory experiments can thus be carried out on the computer. For example, many biological experiments are using techniques in which a certain compound is introduced into the system and perhaps removed or neutralized later. Similarly, it would be desirable to introduce into the model-system various compounds so that their action could be studied. In order to carry out experiments of this nature, various electronic switches were incorporated in the program of the model-system. They were synchronized with computer solutions so that operations could be carried out at any desired time. Also, experiments of a more complex nature were designed so that two compounds could be simultaneously introduced into the system and simultaneously removed from the system. In addition, a third compound could be added subsequently. The transient introduction of various elements into the model-system enabled one not only to follow the events at the site where the interaction took place, but also to follow a sequence of events which took place throughout the system. This is in contrast to experimental procedures in biology where very few entities can be simultaneously studied, because of limitations of experimental techniques. Consequently the amount of information obtained from laboratory experiments can be more limited so far as the total system is concerned. This is especially true for the kinetic aspects of the process. For the purpose of recording, an observation system on the computer was electronically organized so that every entity could be followed throughout the experiment. On all recordings the initial value for the functional entity is indicated by 1, and 0 indicates that the functional entity has disappeared from the system. The value 2 indicates that the particular entity has doubled. Since many experiments were carried out during the study, some general comments seem desirable.

In order to obtain a functional cell growth system, the functional entities have to be in the proper quantitative relation to each other as a function of time. This presupposes that rate constants representing turnover and formation of various functional entities should also be in a proper relation to each other. Basically, we are here dealing with the flow system where continuous synthesis and decay of various elements takes place. If some element grows too fast and another too slow, then finally an imbalance will result and the system will be disorganized. As a matter of fact, in a system containing so many rate constants and variables, the most probable state is a disorganized state. Only by selecting specific rate constant values and initial conditions is it possible to obtain a functional system. Obviously there can be a certain amount of deviation in parameters, but those are limited. For example, it is possible to make some functional entity more unstable by letting it decay faster. This increased loss of entity can be compensated by faster synthesis. However, this procedure already introduces an imbalance in the system because increased synthesis of one entity will affect the other systems, since there is a competitive state between various synthetic processes. This means that if only one element is made more unstable, the whole system has to be altered in order to gain proper balance. During our studies, which lasted several months, some minor changes were occasionally made in the functional system. However, the system was basically always the same as it was initially established. That is, once in a while, in carrying out a particular type of experiment, it was convenient to alter some rate constants in order to fit the experiment into computer observation time scale. Consequently, all experiments are not exactly comparable throughout the long course of study, but they are always comparable in a particular set of studies carried out for particular purposes. In all figures, however, the horizontal axis represents time.

Basically, growth represents a process by which cell mass increases as a function of time. From the experimental point of view, this mass increase can be determined by a variety of

methods.   For example, the cell mass can be determined by
weight or by the increase of optical density.   However, the
determination of intermediate functional elements such as DNA,
RNA, and other specific functional units is not at all simple.
This is because growth represents a dynamic state of the
system where all individual elements are changing as a function
of time and the pattern of change is differing among individual
components.   In order to measure the change of individual
entities as a function of time from the experimental point of
view, it would be essential that measurements be made during
the cell growth cycle, and not merely at the beginning and at
the end of the period of growth. Measurements should be taken
at short, discrete time intervals, so that a rather continuous
function may be recorded.   This process will of course entail
many great difficulties, especially if the cell is very small.
Consequently, in this case measurements have to be carried
out in a mass culture of cells.   An additional difficulty lies
in the fact that the cells are in various phases of division so
that we can obtain only a kind of average data, which does not
actually represent the individual cell growth cycle. Therefore,
various techniques have been developed to get information from
mass cultures.   One of the methods has been the synchroniza-
tion of cell growth in the culture medium.   However, synchro-
nization introduces some metabolic disturbances in the life cycle
of the cell and, consequently, we are not dealing with entirely
normal cell performance.   Nevertheless, with certain precau-
tions much useful information has been obtained by this method
(4).   Other methods have been developed, but no attempt is
made here to discuss such problems because they have been
analyzed widely in the literature.   In contrast, in computer
analysis of model-system, it is easy to follow, during the experiments, the growth of all functional entities. These can be
recorded as a function of time and analyzed in relation to each
other. In our studies a large amount of data has been collected
this way.   However, no attempt is made here to compare very
closely the kinetic data obtained from the cell model system
and the experimental data on cell growth obtained from actual

measurement in the laboratory. In the first place, it is well known that different cells have different types of growth characteristics and these depend, of course, on the dynamic organization of the cell. Secondly, cell growth characteristics have been determined under a variety of environmental conditions and, consequently, there is no basis for directly comparing a general model-system with specific quantitative measurements. However, a general comparison is made throughout our studies between experimental and theoretical data.

4. ANALYSIS OF GROWTH PROCESS. BASIC FUNCTIONAL ENTITIES IN NORMAL GROWTH AS A FUNCTION OF TIME

A computer solution is obtained for the cell growth when the model-system is made operational in the presence of external pool P_e. Since all basic functional entities have the initial value of unity, then the growth will be initiated from that level. A preliminary set of experiments will now be carried out primarily for the purpose of testing the model-system. In order for the system to grow, it is essential that there should be an inflow of pool P_e via transport enzymes E_t. A rate constant value of k_{18} has been so selected that during the generation time all cell components approximately double the initial value. We will observe now, after initiation of growth, the kinetic behavior of various functional entities. Figure 3 shows the growth of three pools P_i, P_a, and P_n. Pool P_i increases rapidly up to a maximum value and then decreases gradually. When observation time is extended beyond generation time, this curve will flatten out and gradually start to rise again. Pool P_a after the initial transient will also start to grow, finally reaching a plateau. It is obvious that the amino acid pool and general internal pool have quite different growth characteristics. In contrast, the nucleotide pool P_n will pass through a steep reduction phase before it starts to grow. This is of course to be expected, since the first step of synthesis will be utilization of P_n by genes, thus causing rapid initial loss of this

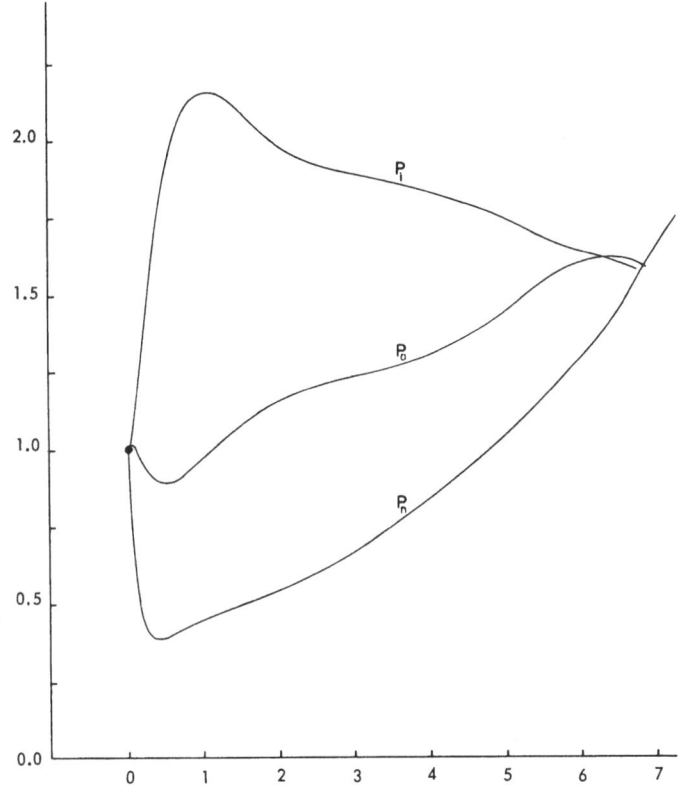

Fig. 3. Concentration of pools P_i, P_a, and P_n during growth.

pool. Figure 4 shows the growth of the enzymes and transport RNA. After an initial transient increase, E_p is reduced, passes through a minimum, and will then grow consistently. In contrast, enzyme E declines immediately, then passes through a minimum, and finally starts to grow. This rapid decline of E is caused principally by a rapid increase of internal pool P_i, since interaction of P_i and E as a regulatory step in the system will cause complex formation (equation 22, Table II), and there is consequently a reduction of free enzyme. However, in general, processes are so complex that no attempt is to be made here to analyze the curve forms specifically.   Transport RNA C

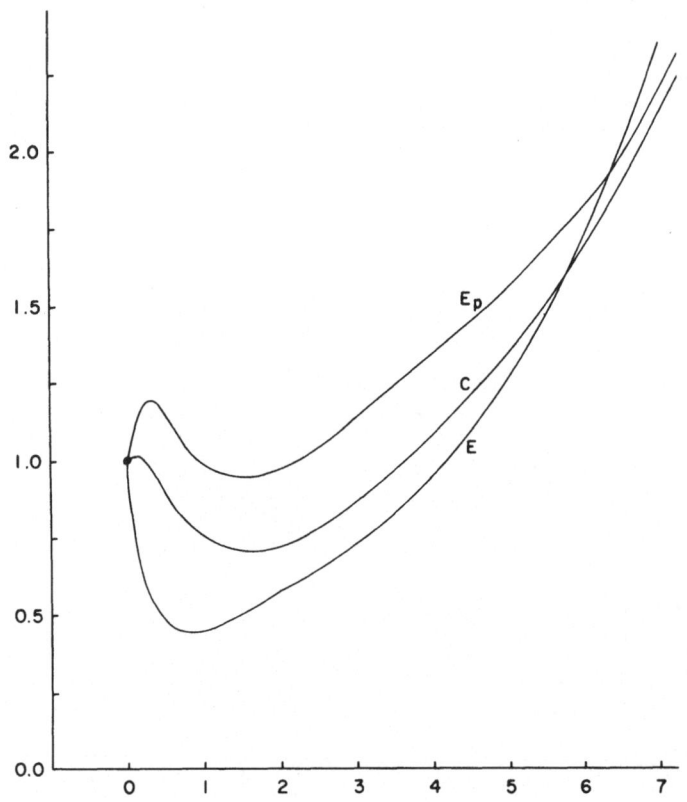

Fig. 4. Concentration of enzymes E_p and E and transport RNA C during growth.

exhibits characteristics which are similar to those of enzyme E_p. Figure 5 shows the formation of messenger M, template N, and ribosome B. They all have different kinetic characteristics. N and B decline initially, while M first increases and then declines. However, they all finally start to grow and reach approximately the same value at the same generation time. Obviously this system would grow and explode if cellular division did not take place. Since we are considering at the present only growth, no attempt is to be made to analyze the growth beyond generation time. This will be done in a later section. In these preliminary model studies, no attempt is made to present all functional entities during growth. These have been recorded, but are excluded in order to avoid repetition, since this material will appear later in the text. In summary, the general kinetic growth characteristics for various functional elements are different. They all start sooner or later to grow and reach double value at the end of generation time. It appears that the growth of the model-system when it starts from the initial values is in an arbitrary state. In the initial phase there are formations of various complexes and adjustment of equilibrium at various steps.   Only after the initial transient phase has passed can we consider the normal growth

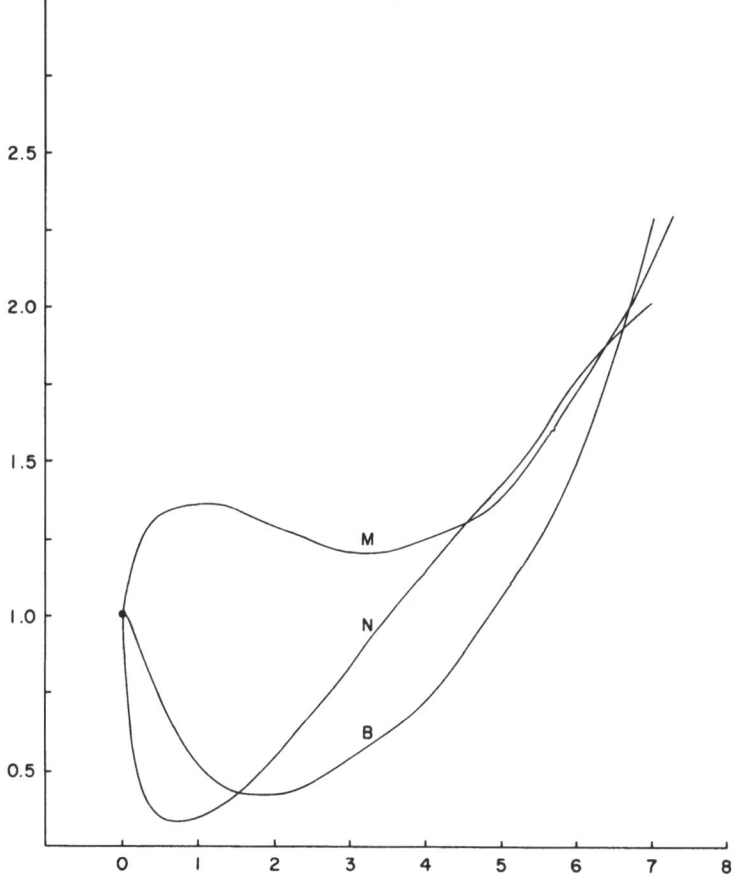

Fig. 5. Concentration of messenger M, template N, and ribosome B during growth.

conditions established and the system to be suitable for kinetic
analysis. In order to minimize the initial transients, one also
has to assign certain initial values for complexes (see
equations 1, 9, 21, 22, and 23, Table II). This is, of course,
extremely difficult to do on a theoretical model. In order to
determine initial complex values, some empirical studies were
carried out. These were determined for conditions where there
are only small initial transients present. However, in the
study of the external pool effect, it was instructive to start the
growth from conditions where intracellular complexes are
absent and there are only functional entities present. Under
these conditions, in the following section we shall carry out
the analysis of pool concentration effect on various functional
entities. The special significance of this analysis lies in the
fact that it will demonstrate whether the model-system has the
ability to organize itself.

a. The Effect of Pool Concentration on Growth Characteristics

We shall first consider growth at various external pool
concentrations. The pool P_e effect can be followed with the
aid of equation 16, Table III. The first term in this equation,
$k_{18} EP_e$, is the generating term. This term represents positive
input into internal pool P_i, while other terms in equation 16 are
turnover terms. At present we do not analyze the turnover
terms and will keep our attention on the first generating term.
It contains rate constant k_{18}, the enzyme E, and pool P_e concentration. Enzyme E is a variable which is involved in the
transport mechanism, and it is not manipulated for pool
transfer analysis. Both k_{18} and P_e are constants during the
experiment, and different values could be assigned during
different experiments. In these experiments P_e concentration
value is varied above and below the normal value which was
established initially for a functional system. We are especially
interested in finding out the critical values at which the functional system becomes nonoperational. Figure 6 shows internal
pool P_i concentration as a function of time at various external
pool P_e concentrations. Numbers on the curves indicate the

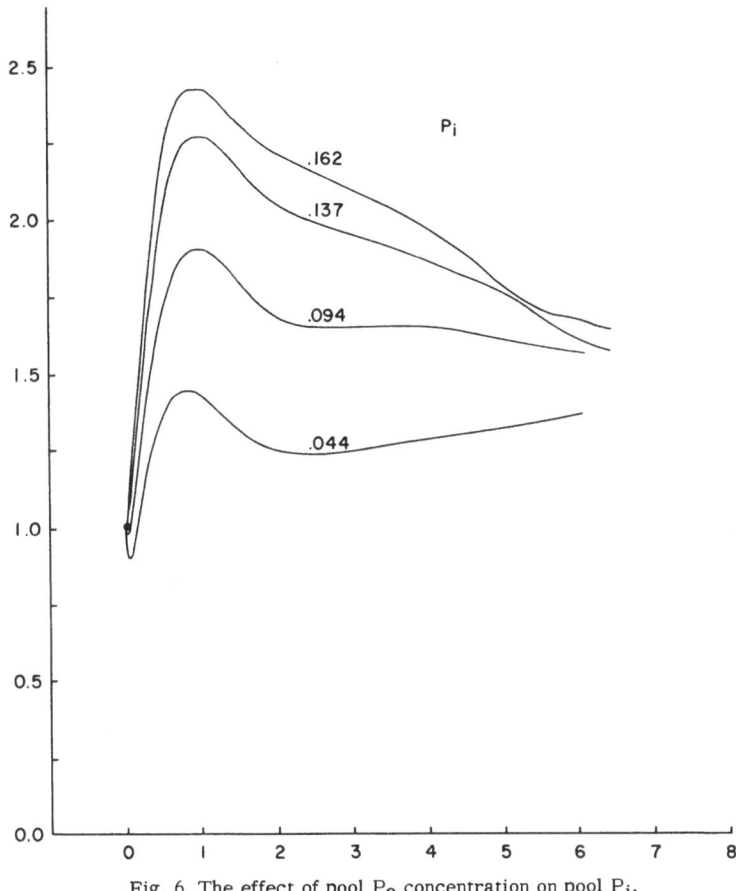

Fig. 6. The effect of pool P_e concentration on pool P_i.

relative values of pool concentrations. The value 0.137 is considered to be normal pool concentration. When pool concentration is increased to 0.162, there is an initial increase of P_i, but at the end of generation time the pool value has returned to approximately the same value as in the normal condition. At the lower P_e value (0.094) P_i also has a lower absolute value, and the curve form flattens out after the initial rise. However, at P_e value of 0.044, considerable changes occur in the basic curve form of P_i. There is an initial reverse gradient followed by the usual increase and decrease. However, here P_i con-

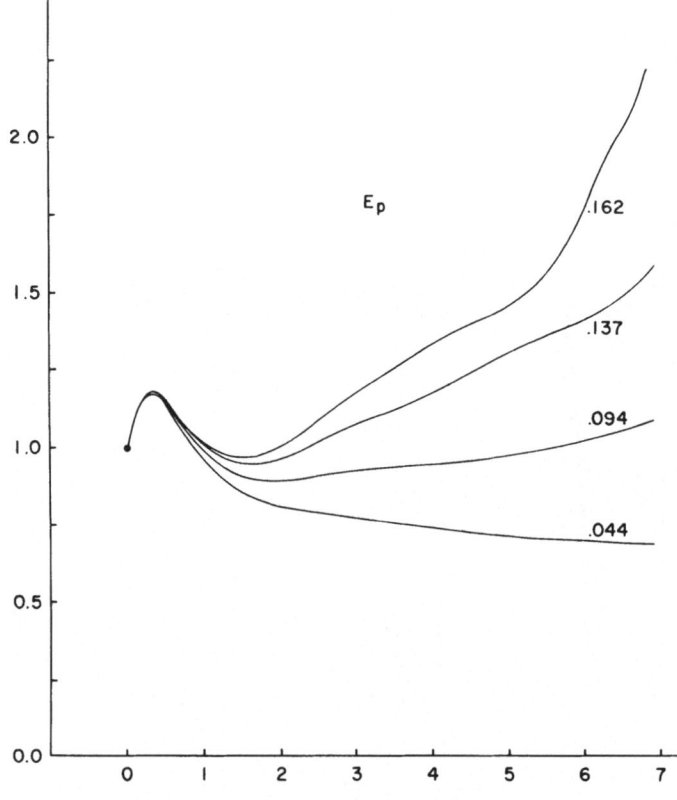

Fig. 7. The effect of pool P_e concentration on enzyme E_p.

centration passes through a minimum and then starts to increase. This is contrary to the behavior of other P_i curves. Figure 7 shows the growth of E_p; it is evident that E_p grows at all external pool values except at P_e value 0.044, where E_p is decreasing in a later phase of generation time. This indicates that the model-system is becoming nonfunctional. This is also evident in the growth of enzyme E, as indicated in Fig. 8, where similar effects are observed. It is interesting to note here that the increase of pool value from 0.137 to 0.162 has very little effect on enzyme formation. A similar effect is also apparent on transport RNA C, as indicated in Fig. 9. Ribosomal RNA B (Fig. 10) grows at all pool P_e values except pool value

0.044, where there is a rather constant ribosome value. Even here, when observation time is extended, ribosomal concentration finally starts to decrease, but it lags behind several other elements.   This is of course to be expected, since the functional stability of ribosomes is higher than that of most of the other components in the system. Messenger M (Fig. 11) grows at all pool P_e values except at the lowest value (0.044).

A general conclusion which can be drawn from these pre-liminary experiments is that the model-system, once organized, is capable of maintaining itself in a condition where the external pool concentration varies within a certain range of values.   When pool P_e concentration falls below a certain critical value, the system will start to decay and finally be-comes nonfunctional. Similar phenomena can also be observed, in cellular biology, where cells are capable of growing only

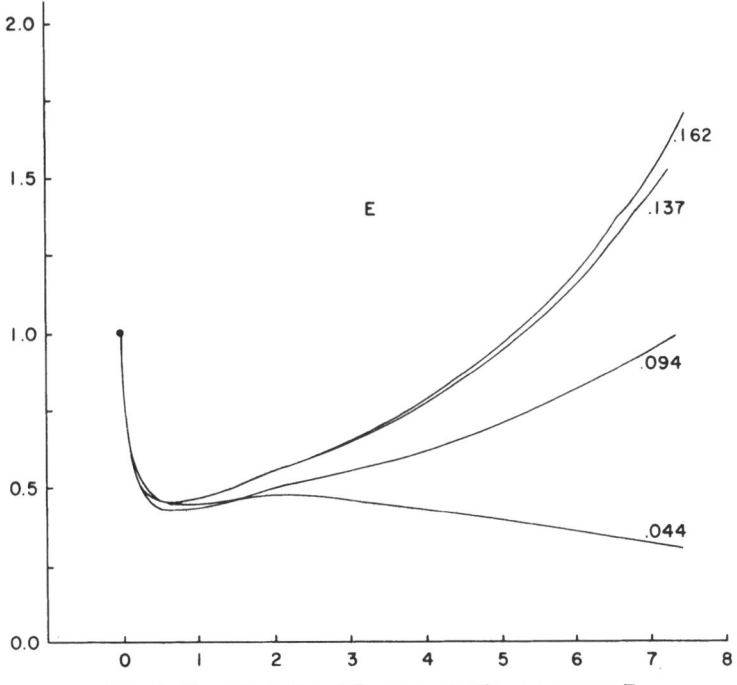

Fig. 8. The effect of pool P_e concentration on enzyme E.

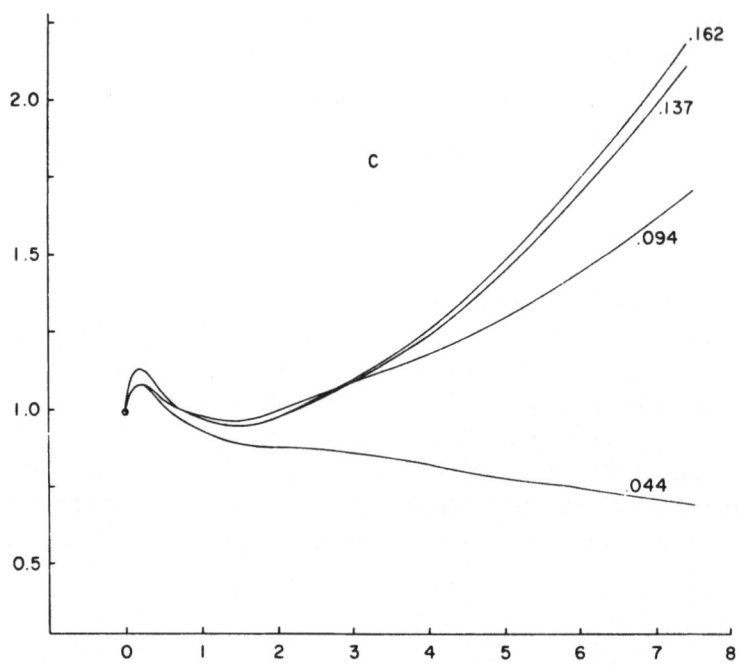

Fig. 9. The effect of pool P_e concentration on transfer RNA C.

above a certain minimal concentration of nutrient medium. In another section of this book we will further analyze the aspect of functional disorganization, but at present we can make the statement that in order to analyze the functional system we have to operate in a certain range of pool values. It should be noted here that the elimination of a part of P_i via rate constant k_{30} into an external medium does not increase P_e. It is considered here that the product is not reutilizable by the system.

b. The Conversion of the Internal Pool into Specific Pool and its Effect on Growth Characteristics

It is well known that the composition of the growth medium as well as the concentration of the medium will affect the cell growth rate as well as the generation time. It would be interesting to follow the experiments on the computer and observe how the conversion of the internal pool P_i will affect the growth rate and growth characteristics when conversion rate is varied. We consider here two pools, P_n, and P_a, and analyze them separately. The rate of change of pool P_n concentration is presented in Table III, equation 2. It is evident that the first term, $k_{19}EP_i$, is the only generating term for increase of cell mass, while other positive terms, $k_{12}M$ and $k_{13}M_p$, represent the messenger decomposition, and thus cause only a turnover of P_n. Pool utilization is represented by the rest of the terms. Rate constant k_{19} represents a composite term for multiple-step processes, containing nucleotides and subunits. The conversion process could be considered faster when pool P_i composition is such that it contains precursors and molecules which can be used in the final assembly of RNA. Consequently, an internal pool, P_i, which has a wide spectrum

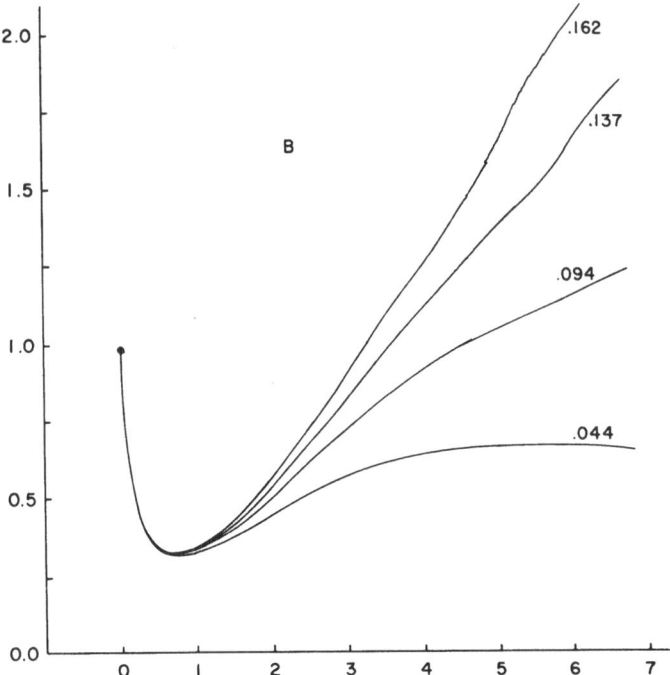

Fig. 10. The effect of pool P_e concentration on ribosome B.

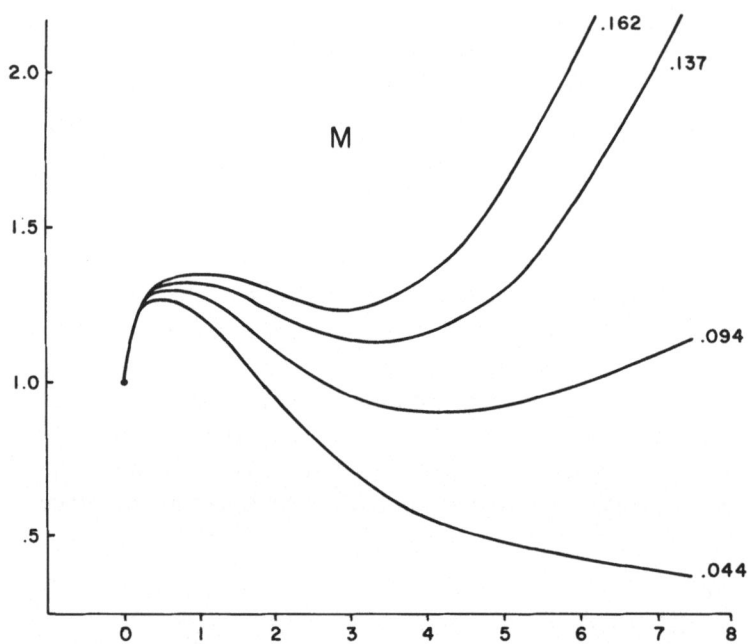

Fig. 11. The effect of pool P_e concentration on messenger M.

of precursor components for pool P_n, can be considered to turn over faster and when the system has a poor precursor distribution, it will take a longer time to establish final pool. It would be of interest to study how the rate of nucleotide pool formation will affect growth characteristics of various functional entities. For this purpose the rate constant k_{19} is varied within a wide range and other entities are recorded. It is of interest to find out first what happens to pool P_n when rate constant k_{19} is varied. Data are presented in Fig. 12, where numbers on curves indicate relative k_{19} values in different experiments. When other functional entities are recorded, the same k_{19} values are also represented on various curves all through this section. When k_{19} has relative value 3, which is the lowest value recorded here, then there is initially a very rapid reduction of P_n, and after passing through a minimum it will start to increase slowly. When the value of k_{19} is increased to 4, then the initial drop of P_n concentration is smaller than in previous experiments and the growth of the pool is faster. This phenomenon is more pronounced when the k_{19} value is increased to 7.5. However, when the rate constant is increased to 11, there is a lesser reduction in the early phase, but subsequent growth is reduced. A drastic reduction in P_n concentration occurs when k_{19} has a value of 15. Initial pool reduction is here the smallest, but later growth is comparable to the condition when k_{19} has value of 3. It is evident that excessive conversion of P_i to P_n limits growth, and there seems to be an optimum k_{19} value which gives the highest growth rate for this particular system. No attempt has been made to evaluate the optimum value for k_{19} because its numerical value has no general significance. However, the phenomenon itself is highly significant, showing that excessive concentration of metabolities may have inhibitory effects on growth.

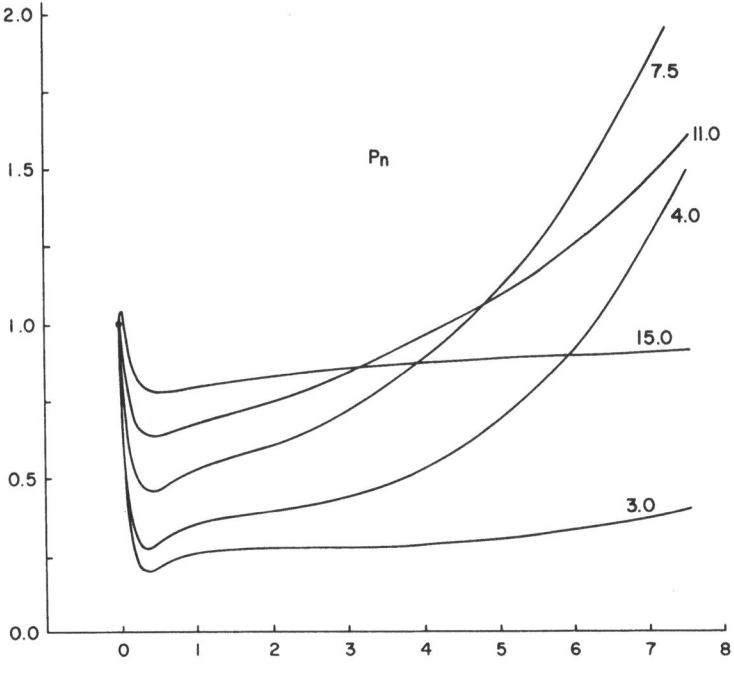

Fig. 12. The effect of rate constant k_{19} on pool P_n.

Figure 13 shows the effect of k_{19} changes on amino acid
pool P_a. The effects here are the opposite of those observed
with P_n. At the lowest value ($k_{19} = 3$) there is the largest
increase of P_a. An increase of k_{19} decreases the P_a concentration. Such a relation is not unexpected, since both P_a
and P_n compete for the internal pool P_i. In an analogy with
cellular growth, one must consider that while P_a and P_n have
their own specific compositions, there are some general elements in the pool P_i which are shared by both pools. It is
evident in Figure 13 that high k_{19} values will suppress the P_a
growth. The complexity of various interactions between the
functional entities within the model-system is well demonstrated
in Fig. 14, where transport RNA C growth is presented. It is
evident that at low values of k_{19} (3 and 4), there is initially a
rapid decline; the curve then passes through a minimum,

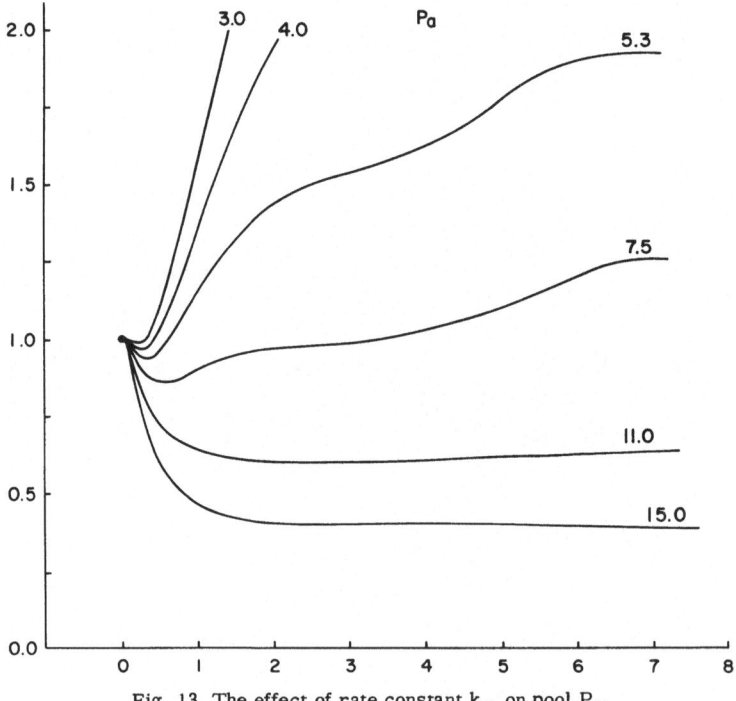

Fig. 13. The effect of rate constant k_{19} on pool P_a.

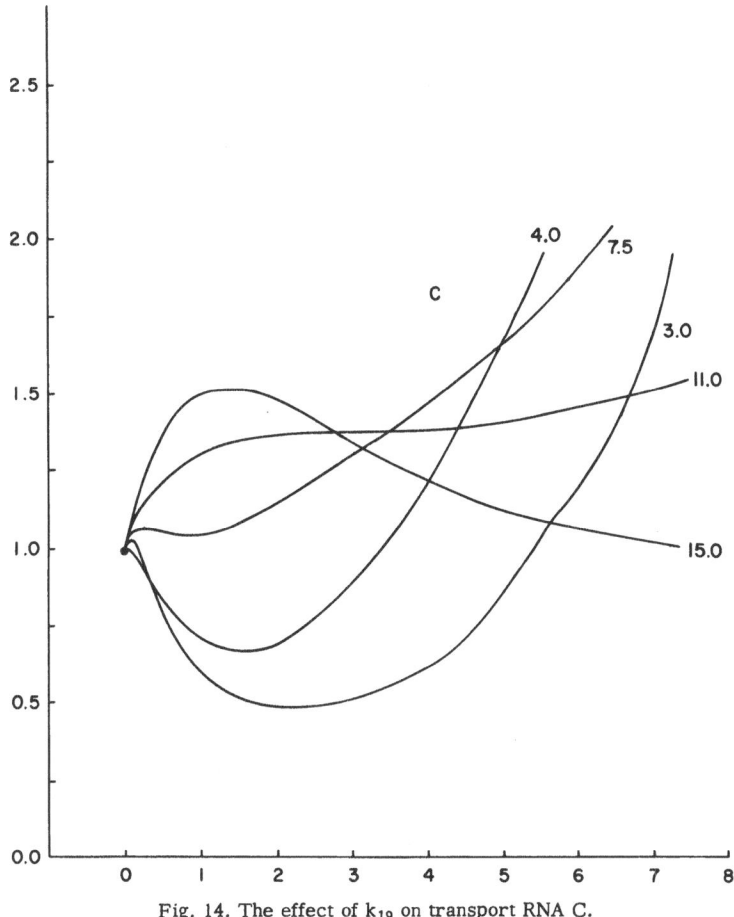

Fig. 14. The effect of k_{19} on transport RNA C.

and C will gradually begin to grow rapidly. At a k_{19} value of 7.5, there is a small initial rise followed by smooth growth. When k_{19} is increased to 11, there is an initial rise, followed by slow growth. However, when the k_{19} value is 15, a relatively rapid increase takes place initially, reaching a maximun, after which a steady decline of C concentration takes place. These experiments clearly demonstrate that it is impossible to predict by casual reasoning how complex systems behave in various conditions. Since ribosome concentration has

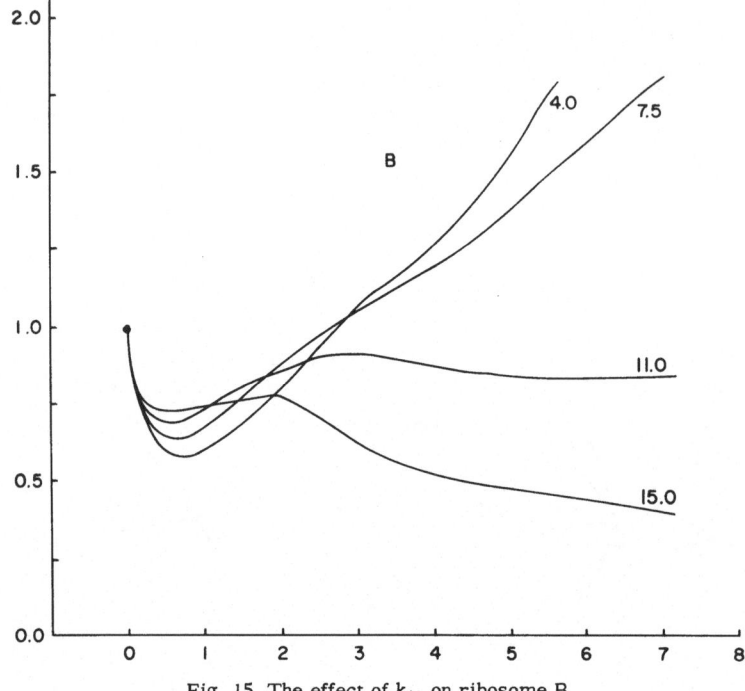

Fig. 15. The effect of k_{19} on ribosome B.

usually been linked with cell growth, it is of special interest to
follow its growth characteristics in relation to pool P_n.
Figure 15 shows the growth of the ribosomes at various k_{19}
values.   At intermediate k_{19} values (4 and 7.5), there is rapid
growth.   However, in contrast, high k_{19} values are associated
with the reduction of ribosome concentration. The kinetics of
such a system is obviously too complex to pinpoint the principal
factors for development of such growth dynamics. One aspect
of organizational characteristics which contributes to complex
relations is the specific structural feature of ribosomes
containing both protein and RNA.   It is of particular interest
to observe what happens to enzyme synthesis in this type of
condition.   It appears from Fig. 16 that enzyme E synthesis
is fastest when k_{19} values are low, in which case pool P_a is
high.   Here again at high k_{19} values there is very slow growth
($k_{19} = 11.0$) or there is a reduction of growth ($k_{19} = 15.0$).

Fig. 17 shows that polymerase E_p, and the k_{19} effect here are generally the same as in Figure 16 for E. The general conclusion which can be drawn here is that excessive values of k_{19}, either too large or too small, will produce abnormal relations between functional entities. It appears that at the end of generation time some components grow much more than double in value and other components grow hardly at all.

Consequently, in order to have a proper functional system, it is essential that the pool P_n have values only within a certain range of concentration. Extreme rates of P_i conversion to P_n will disorganize the orderly growth.

Next we consider the rate effects arising from the conversion of pool P_i into amino acid pool P_a. Table III, equation 13, represents the rate change of pool P_a. It is evident that the main generating term $k_{20}P_iE$ is balanced only by the opposite

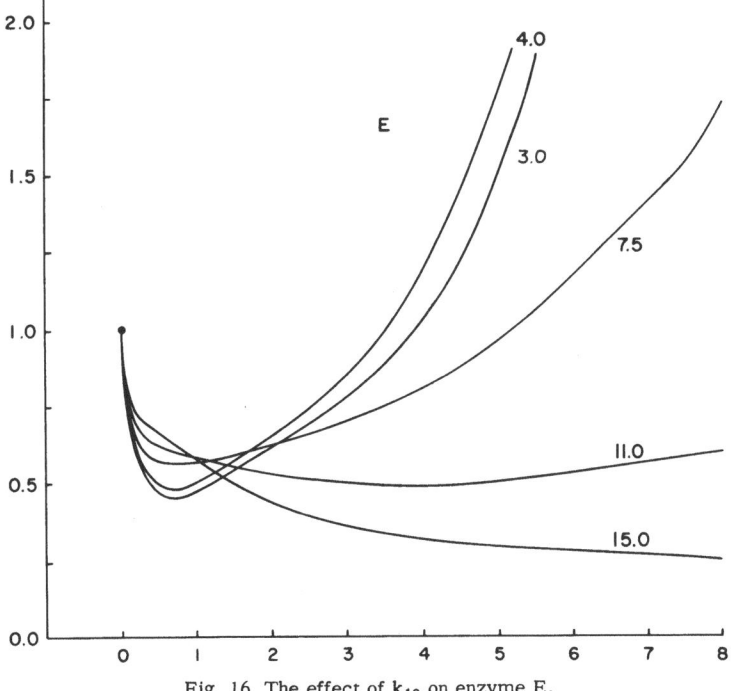

Fig. 16. The effect of k_{19} on enzyme E.

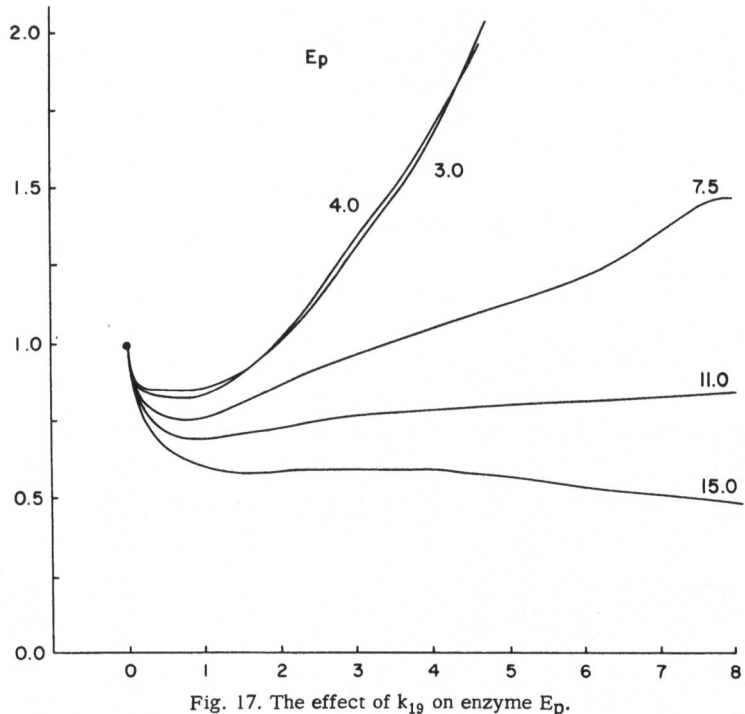

Fig. 17. The effect of k_{19} on enzyme E_p.

term $k_9 C P_a$, which represents utilization of the pool in enzyme
synthesis. In the model-system, enzymes are considered to
decompose into P_i. The transfer of internal pool P_i into P_a is
accomplished via rate constant k_{20}. Figure 18 shows the
concentration of internal pool P_i at various k_{20} values. It is
evident that higher k_{20} values produce more rapid growth and
that low values will terminate growth. It is of interest here
that growth of pool P_n (Fig. 19a), which is a competitor to
P_a, shows in general the same characteristics as the internal
pool, P_i. Figures 21a–23a represent the growth of various
functional entities such as ribosome B, messenger M, and
enzyme E. They all have for the most part the same growth
trend, namely, at low values there is no growth or termination
of growth, while at high values of k_{20} the growth is more rapid.
The only exception is an irregularity occurring in transport
RNA C during growth at low k_{20} values (Fig. 20a). It is of

interest to note here that no growth saturation occurs at high P_a values. The question is, what are the limiting factors for the growth in these conditions   Starting from pool P_a, a logical sequence of events takes place. It appears that P_a is a principal element required for the growth of enzymes: enzymes increase the pools, and pools produce new enzymes and also elements of RNA. There thus seems to be a progressive growth going on, except at the saturation limits of genetic reactions, since these are the only elements which do not grow. In a viable cell several other elements exist which will not be considered here, such as the surface-volume ratio of the cell or diffusion,

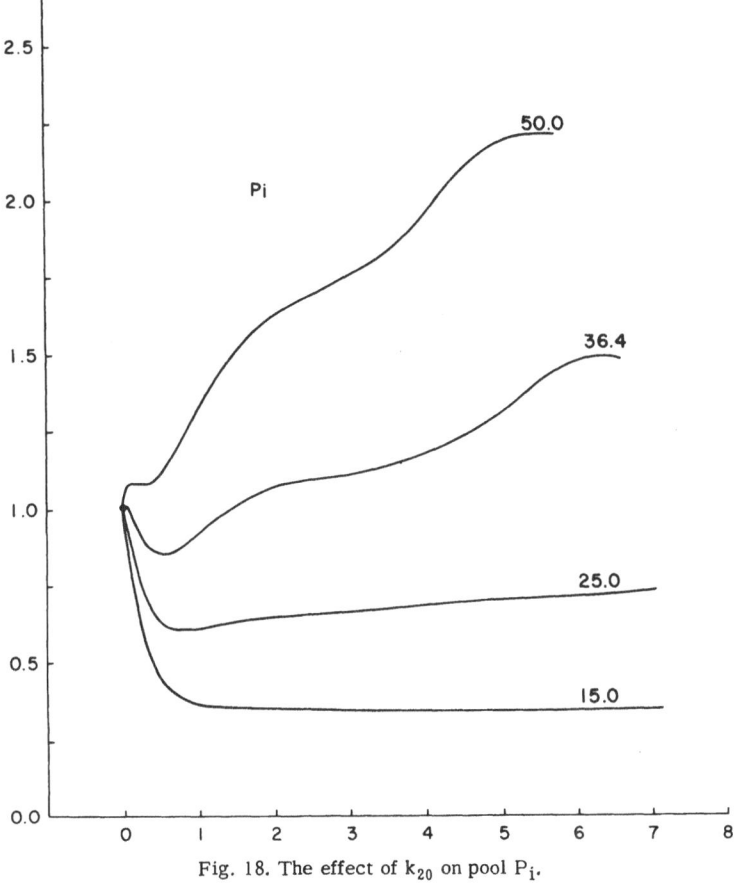

Fig. 18. The effect of k_{20} on pool P_i.

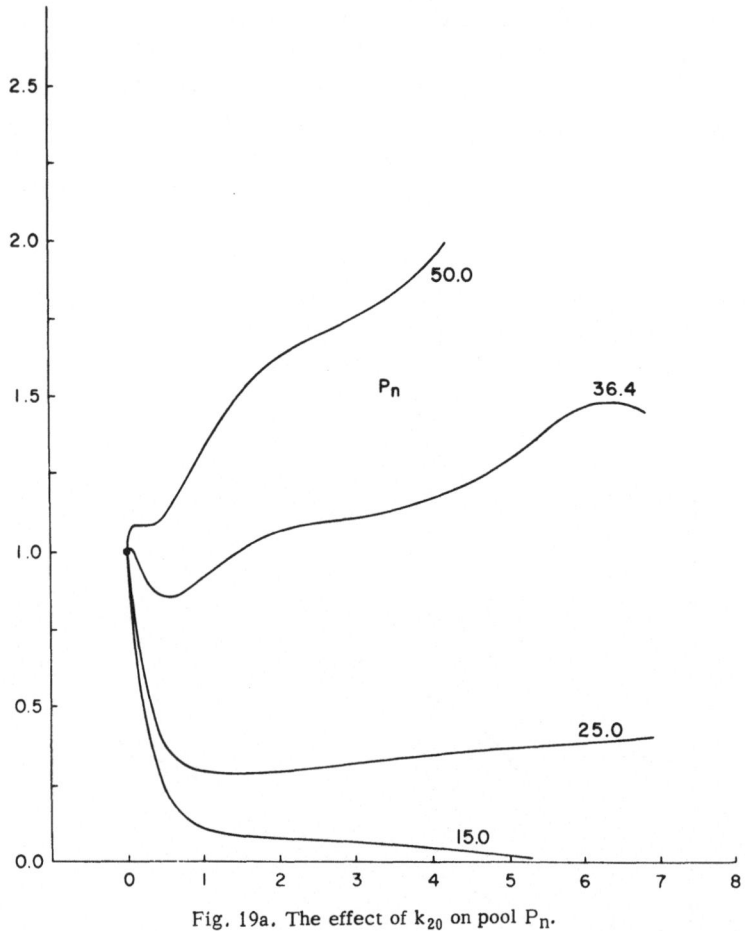

Fig. 19a. The effect of k_{20} on pool P_n.

which assumes importance when the system becomes much larger than what we consider a normal, average-sized cell produced during a generation.

The model-system does not contain such features. These experiments suggest that at the level of protein synthesis a very effective control can be exercised on total cell growth. Furthermore, a rather uniform response can be obtained on the level of all functional entities, which is essential for organized growth. The same cannot be said for control of growth via RNA pool P_n. Since we are dealing here with a model-system, one cannot be too dogmatic in drawing conclusions. But it appears that a model-system as presented has a strong all-over cross-behavior of an efficiently self-organizing system. Consequently, these simulation experiments suggest that a smooth operational control of cell growth seems to be indicated also on the protein level in actual cell growth. It was of particular interest that the cell growth model was able to carry out a growth process in conditions which were not at all optimum for its performance, i.e., conditions where

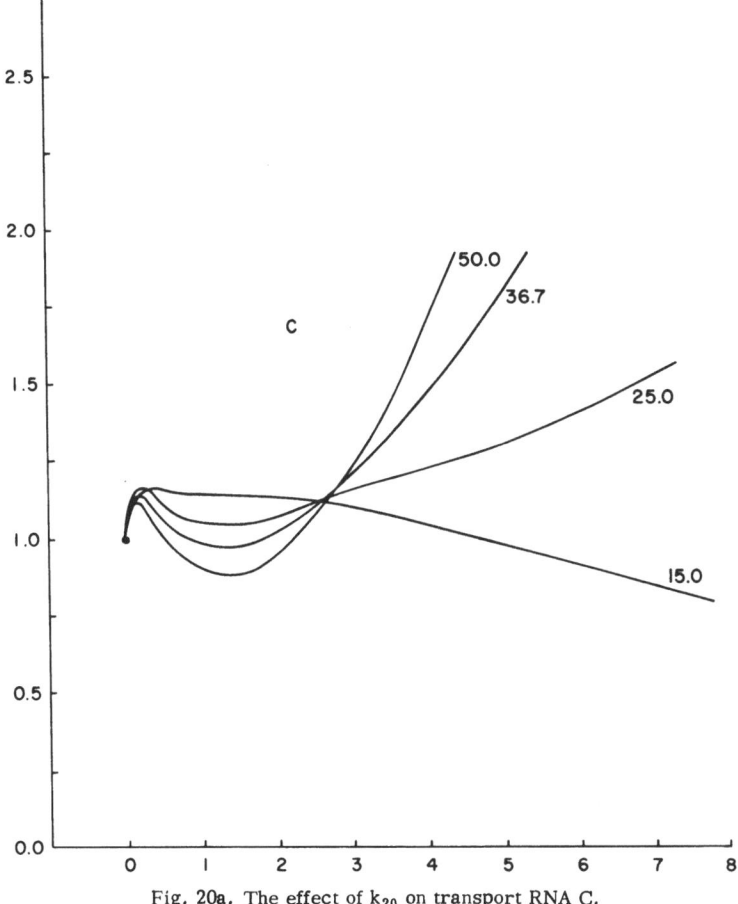

Fig. 20a. The effect of k_{20} on transport RNA C.

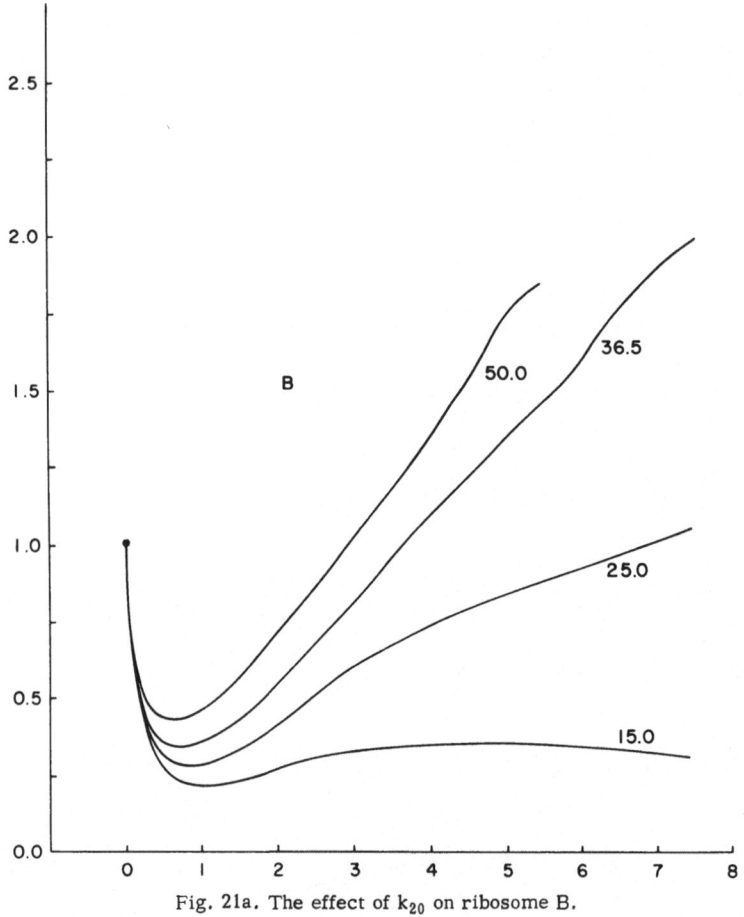

Fig. 21a. The effect of k_{20} on ribosome B.

the intermediate functional elements (complexes) were not
adjusted for optimal performance. The model-system was
still able to organize itself for growth. This indicates that
the model-system has self-organizing characteristcs which are
rather strong, and that it is able to adapt itself when transferred
from one condition to another. However, since we were interested in studying the growth characteristics of individual functional elements in a more detailed manner in conditions which
alter very rapidly, it was desirable that a further parameter
adjustment be made, in order to reduce initial transients and

increase the useful observation time. This was carried out empirically. Such features are especially important in experiments where cell growth conditions change very rapidly. These are typical for so-called shift-up and shift-down nutritional experiments [3,4].

c. Cell Growth in Constant Nutrient Pool

After the reorganization of the model-system, it would be of interest to follow through the growth characteristics of the principal functional entities in the condition of normal growth.

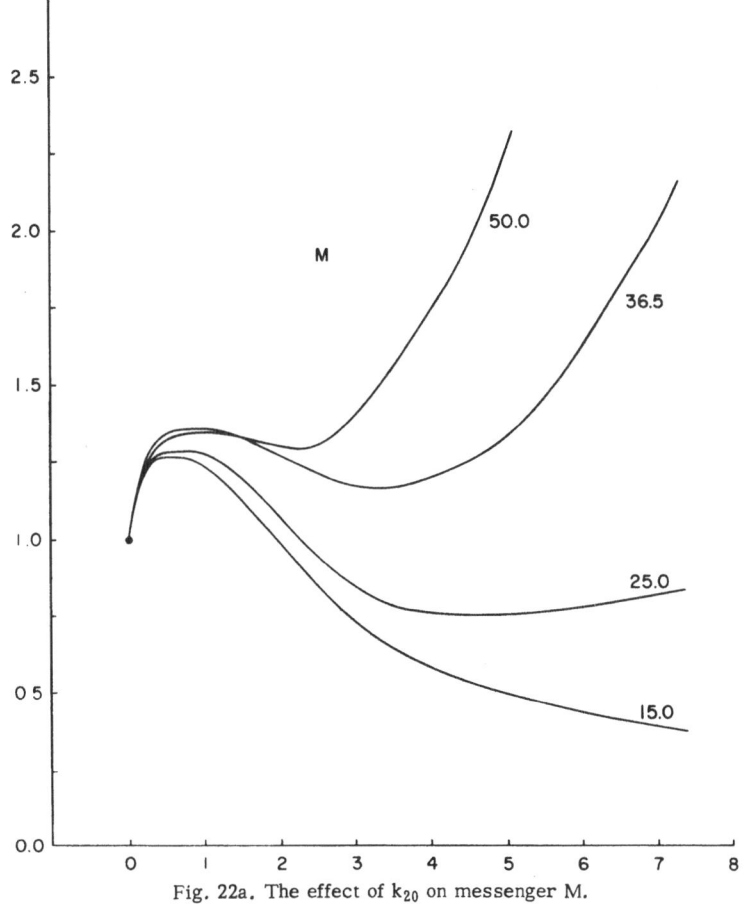

Fig. 22a. The effect of k_{20} on messenger M.

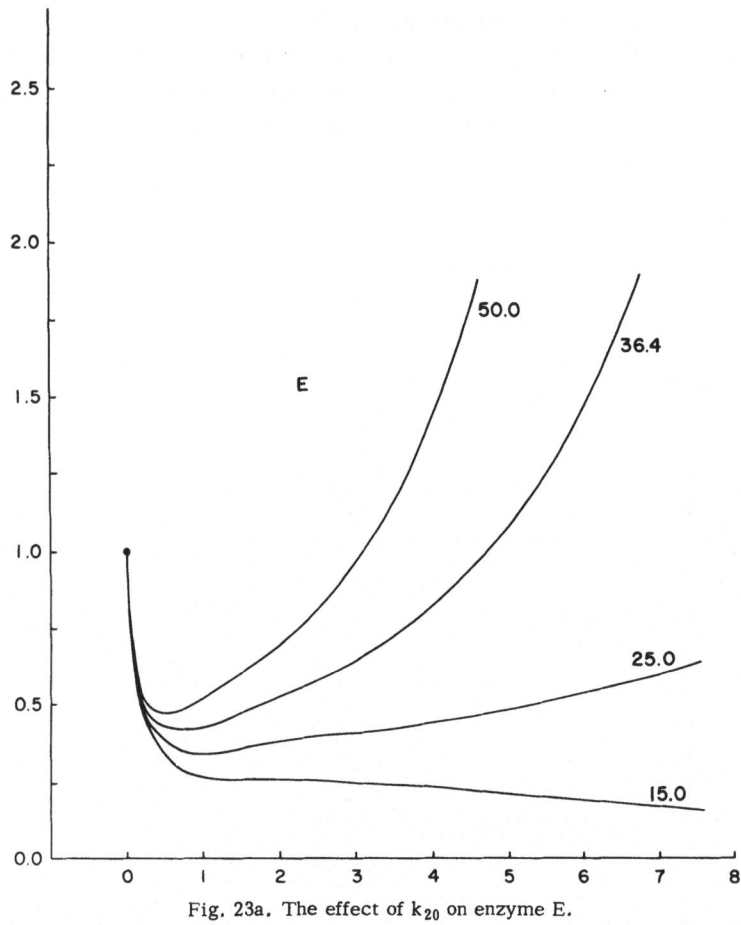

Fig. 23a. The effect of k_{20} on enzyme E.

What is the sequence of the events when a cell model is made operational in the presence of pool? Since an active enzyme system is present, as represented by initial conditions, obviously the first growth step is a transfer of external pool P_e into internal pool P_i. First it would be of interest to see how the pool P_i behaves kinetically during one generation cycle. This pool analysis has been carried out to some degree in the previous section, but here we make some detailed comments on the kinetics of growth. Figure 19b shows the inter-relationship between internal pools P_i, P_a, and P_n and

enzyme E synthesis.   While direct comparative analysis
between the model-system and a growing cell is not feasible,
it would nevertheless perhaps be worthwhile to consider
whether there is any similarity between kinetics of the model-
system and some experimental entity measured during cell
growth.   There are extremely few accurate studies in the
literature which analyze in detail the growth kinetics of
individual functional entities.   Perhaps the most interesting is
the determination of cellular components during the growing

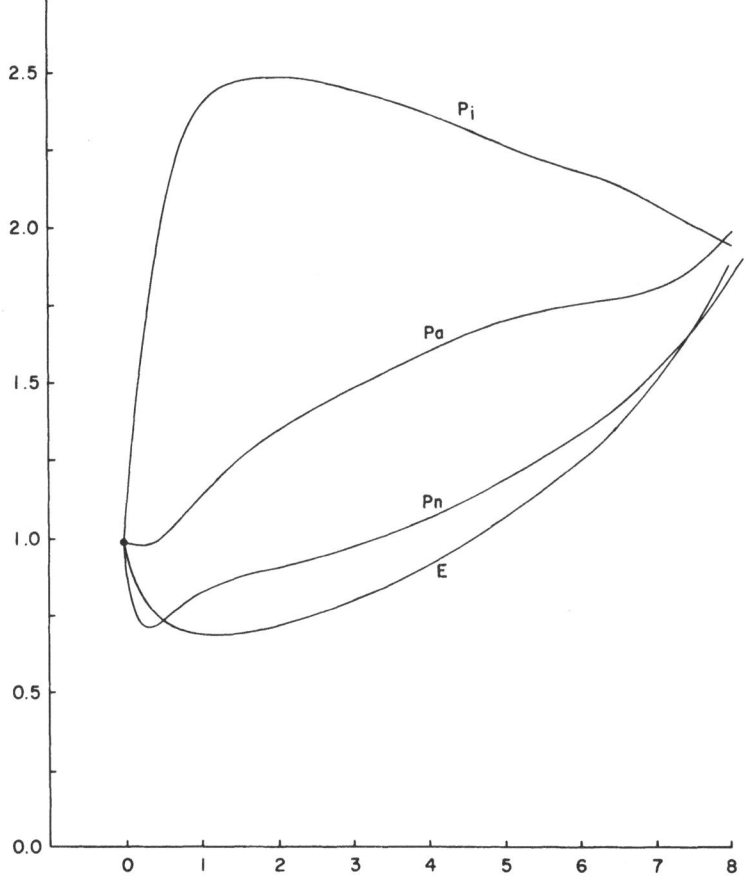

Fig. 19b. The concentration of enzyme E and pools P_i, P_a, and P_n during normal
growth.

cycle in the yeast cells [5]. In these studies the acid extract-able pool starts initially to increase, reaches a maximum, and then declines during the second part of generation time. The curve P_i in Fig. 19b exhibits roughly similar characteristics. However, model-system studies indicate that P_i characteristics can be modified by varying the various parameters and initial conditions of the system. In relation to P_i, the pool P_a starts to grow much slower, and pool P_n passes through a deep initial decline which is subsequently followed by steady growth. This

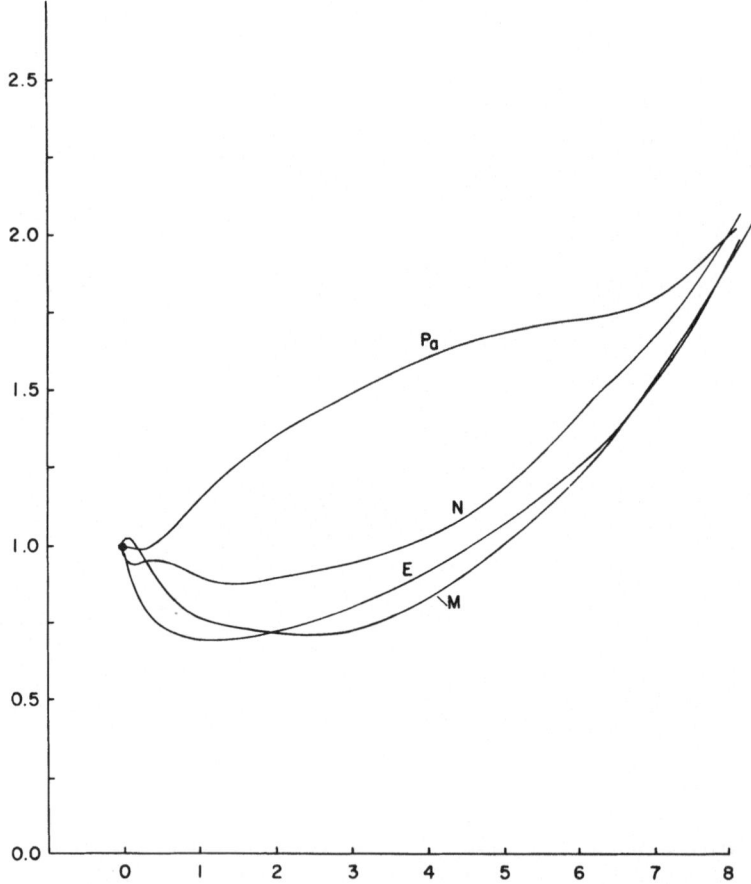

Fig. 20b. Concentration of pool P_a, template N, enzyme E, and messenger M during normal growth.

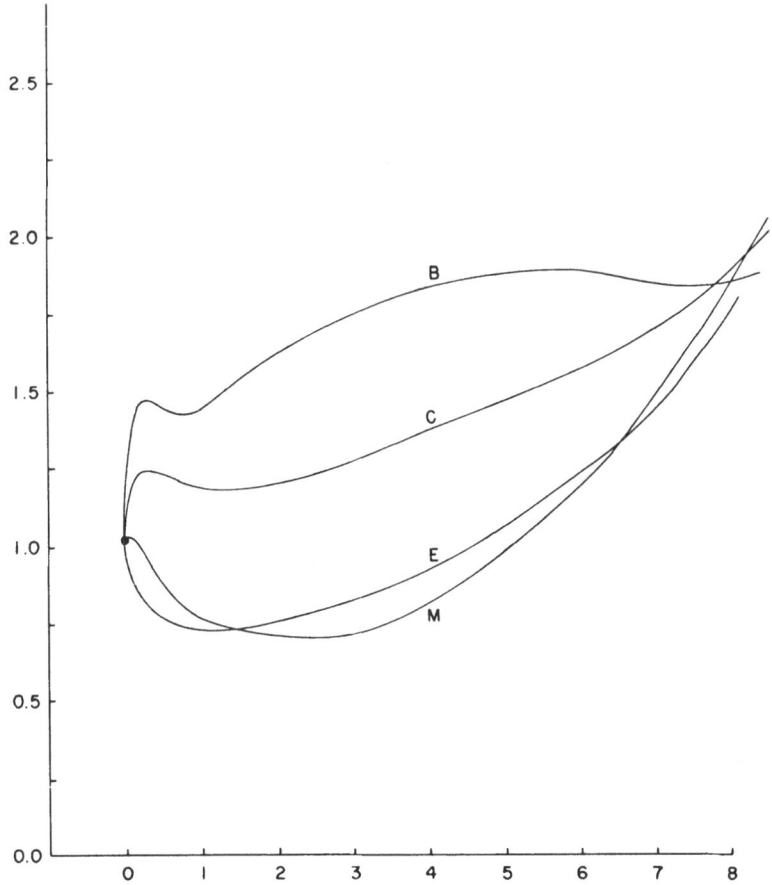

Fig. 21b. Concentration of ribosome B, transport RNA C, enzyme E, and messenger M during normal growth.

first decline of P_n is caused by rapid RNA synthesis in the initial phase of growth.

The sequence of events leading to enzyme synthesis is as follows: first messenger M forms followed by the formation of template N by the complexing of ribosome B with messenger M; then pool P_a interacts with transport RNA C, to yield [C P_a] complex, which interacts with N, thereby completing enzyme synthesis. In Table III, equation 15 gives the rate of change in enzyme concentration. Here the complex [EP_i] acts as a

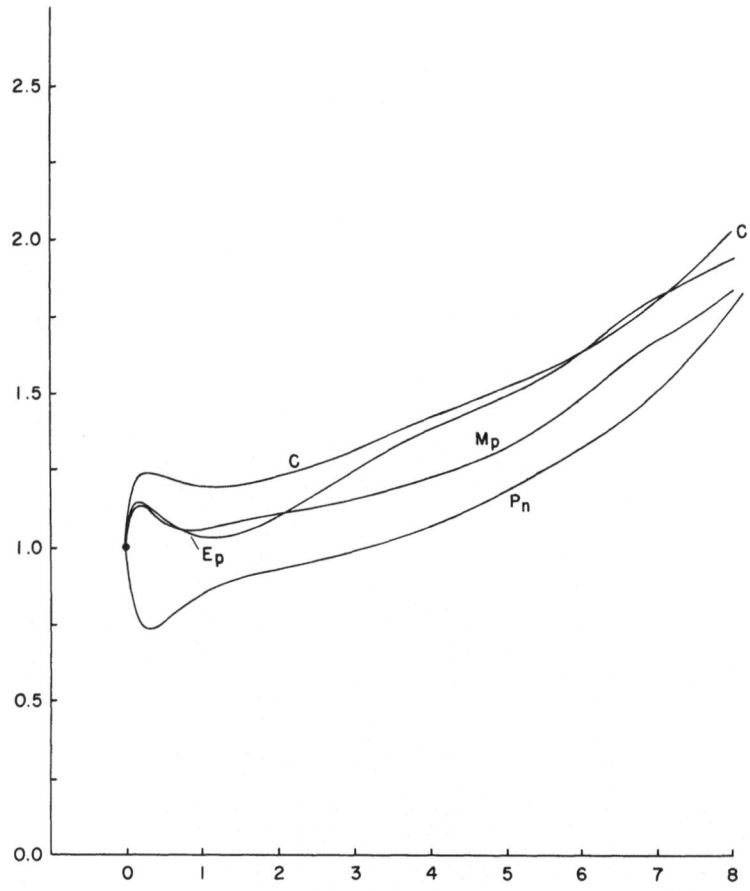

Fig. 22b. Concentration of transport RNA C, enzyme E_p, messenger M_p, and pool
P_n during normal growth.

regulatory mechanism and reduces the free enzyme concentration. In Fig. 20b, the interrelationship between messenger M,
template N, pool P_a, and enzyme E is revealed. It is evident
that after the initial transient the synthesis of M, N, and E is
closely related kinetically. The interrelationship between
ribosome B, transfer RNA C, enzyme E, and messenger M is
indicated in Fig. 21b. It is evident that ribosome concentration
reaches a maximum value and levels out. However, observation

on a more extended time scale shows that B starts to increase again. Kinetic characteristics of polymerase E_p, messenger M_p, and transport RNA C are revealed in Fig. 22b. It appears that all functional entities during the growth cycle increase while they exhibit some degree of different kinetic behavior. A more detailed kinetic analysis is not worthwhile, since the differential equations in Table III reveal that a multiple set of interactions exists between all functional elements and specific analysis of those effects is rather futile in view of the complexity of the system. Computer experiments also show that by altering the various rate constants, different growth characteristics can be produced by various functional entities.

d. The Effects of Transients in External Pool Concentration

In a living system there is a very high degree of integration and regulation between all functional entities during growth. In order to clarify these interrelationships, various experimental approaches have been tried. One line of study involves the determination of relationships among the various functional entities, as well as the sequence of events taking place when cell growth conditions are altered. For example, some experiments have been carried out in which sudden changes are made in nutrient medium during cellular growth [4, 5]. It is evident that the change in conditions of growth also produces changes in protein and RNA concentration. Furthermore, it appears that certain regulatory mechanisms exercise control over these processes. It would be of interest to follow through some experiments on a model-system and see how the transient pool conditions influence the growth.

e. Shift-Down Experiments

To simulate shift-down experiments on the computer the external pool P_e concentration is suddenly reduced to a certain value during a normal phase of growth. This has been accomplished by synchronizing certain electronic relay mechanisms which control the concentration of P_e with the general solution.

Then, rapid P_e concentration changes can be made at any desired time, and the transition of the solution may be studied on the computer under the new conditions. This would be equivalent to experimental conditions in cell growth where the nutrient medium was suddenly diluted and growth changes followed. In order to gain more insight into the sequence of various interactions in the model-system, especially in the behavior of regulatory mechanisms during drastic changes in growth conditions, the external pool concentration was varied. In these studies all entities of the model-system were followed on the analog computer. However, for the sake of avoiding excessive presentation of graphic material, only the principal elements are presented here.

The effect of pool concentration reduction from the value $P_e = 1.0$ to the value $P_e = 0.8$ is shown in Fig. 23. At the time indicated by the arrow, the pool concentration change took place. It is evident that after the reduction of external pool P_e, internal pool P_i is also immediately reduced. In order to evaluate changes in the growth kinetics, these curves should be compared with the normal growth curves, which have been presented previously (Figs. 19b–22b). Figure 23 shows that under these conditions the growth rate of ribosomes, transport RNA, and messenger M is reduced. However, a significant change occurs in enzyme E concentration, where there is an initial increase in the growth rate instead of a decrease. It appears, however, that a small reduction of external pool P_e does not produce any drastic changes in growth characteristics. Figure 24 shows the interrelationship between functional elements when the external pool is reduced to value $P_e = 0.6$ at the time indicated by the arrow. It is evident that a drastic reduction of pool P_i concentration takes place. There is also a marked increase of enzyme E concentration. This is a result of the operation of a regulatory mechanism (equation 15, Table III). Here, the reduction of the P_i concentration shifts the equilibrium, and enzyme E is liberated from the complex $[P_iE]$. Consequently, the regulation of enzyme concentration is effectively compensated when pool P_i is reduced. It is

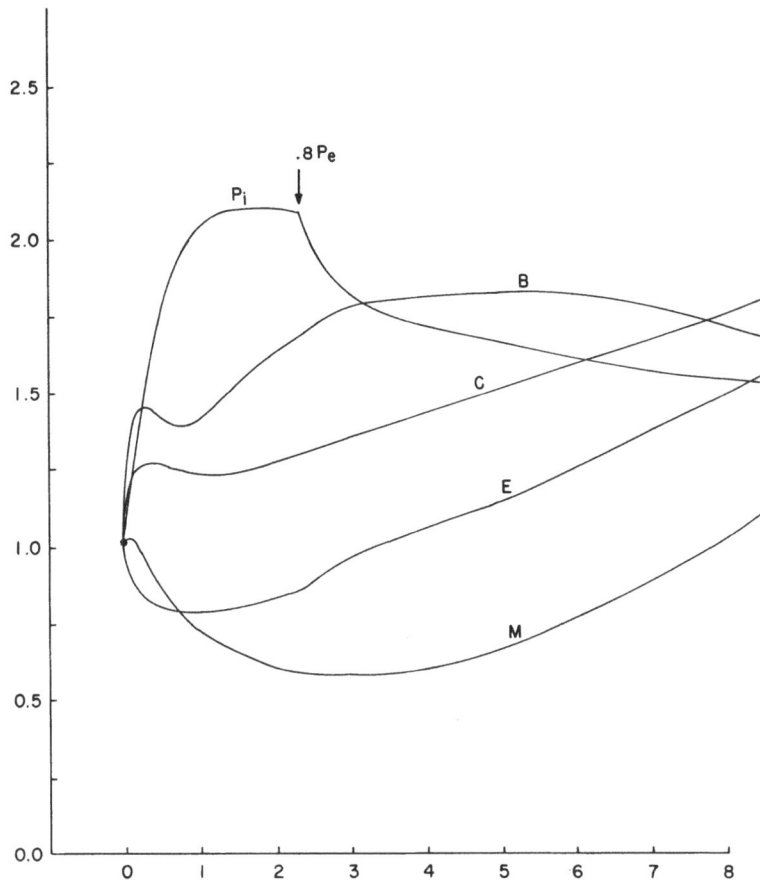

Fig. 23. Shift-down experiment. At the time indicated by arrow, pool P_e is reduced 20%. The effect on: Ribosome B, transport RNA C, enzyme E, messenger M, and pool P_i.

indeed evident that the regulatory mechanism of the model-system functions properly when external changes occur. Of course, this enzyme concentration increase is only temporary. After it reaches a certain maximum value, there is a slow decrease. This reduction of enzyme E concentration seems to be the result of the two following events. First, messenger M reveals a phase of slow decrease instead of an increase, but M finally, after passing through a minimum, exhibits a

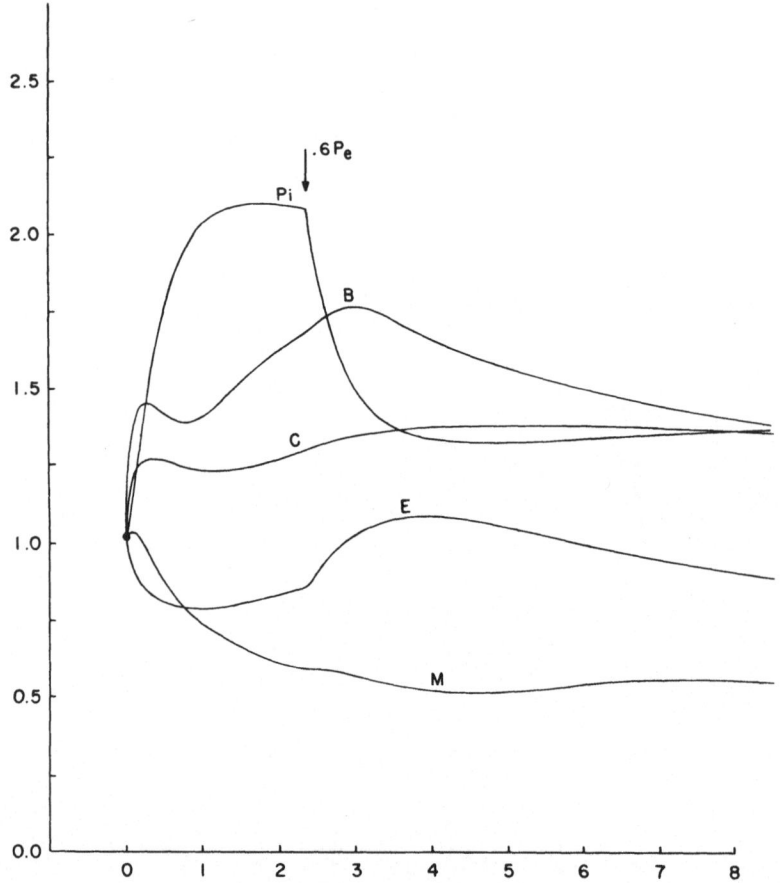

Fig. 24. Shift–down experiment. At the time indicated by arrow, pool P_e is reduced 40%. The effect on: Ribosome B, transport RNA C, enzyme E, messenger M, and pool P_i.

slight increase. Secondly, ribosome B concentration, which was increasing shortly after the introduction of the reduced pool, will cease to grow and will go through a maximum and subsequently decrease slowly. Ribosome B and messenger M concentration reduction coupled with pool P_a reduction are the principal contributors to the reduced enzyme synthesis. It appears that the model–system at this pool level ($P_e = 0.6$) gradually adjust itself to a low level of growth. This is revealed by observation of all functional entities in an extended time scale on the computer.

In order to see if this functional system is able to maintain its characteristics when pool P_e concentration is further reduced, an additional experiment was carried out. The data are exhibited in Figs. 25 and 26, where the pool concentration was reduced to the value $P_e = 0.4$. In these experiments additional entities were recorded, so that it would be possible to

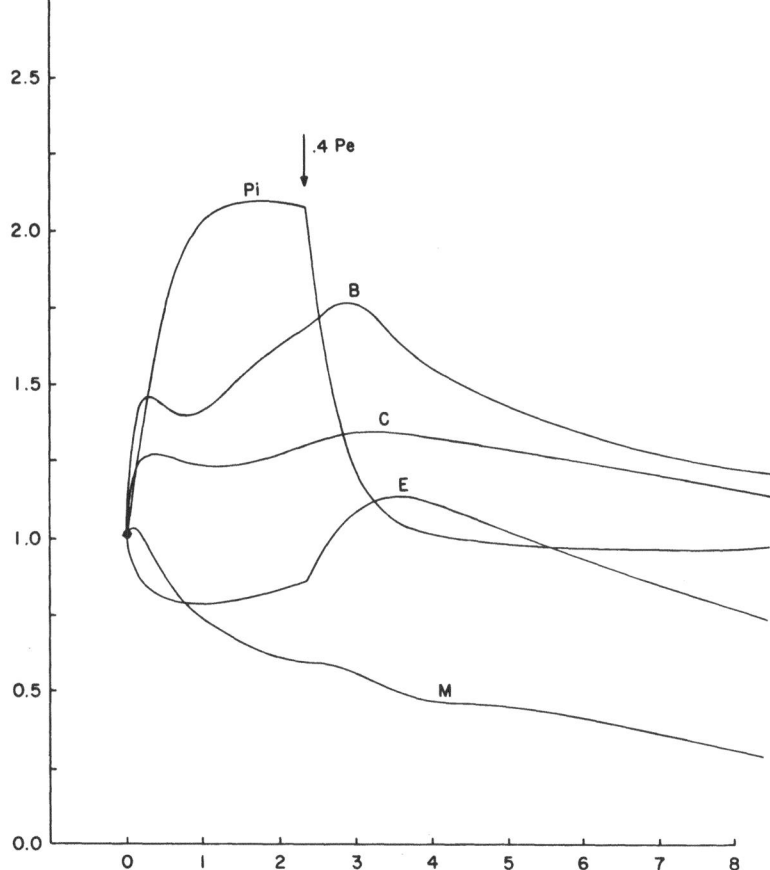

Fig. 25. Shift-down experiment. At time indicated by arrow, pool P_e is reduced 60%. The effect on: Ribosome B, transport RNA C, enzyme E, messenger M, and pool P_i.

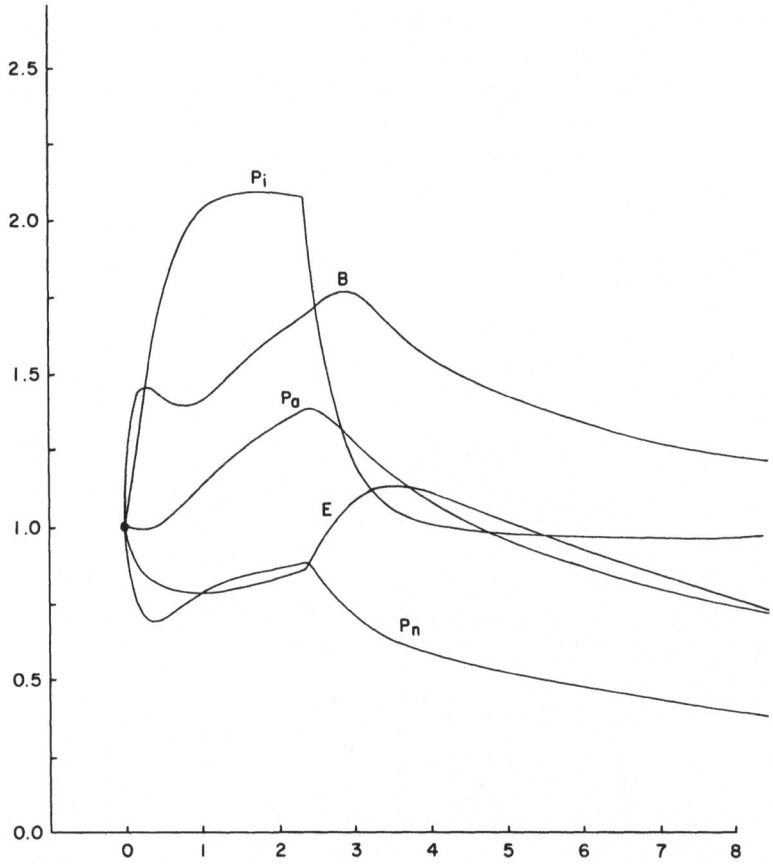

Fig. 26. Same shift-down experiment as in Fig. 25. The effect on: Pools P_a and P_n.

analyze the interrelationships and sequence of events taking place during such drastic pool changes. Here again the reduction of pool P_e is associated with a very marked reduction of pool P_i, and this is associated with a rapid increase of free enzyme E concentration. It is evident from Fig. 25 that the concentrations of ribosome B, transport RNA C, and enzyme E all pass through a maximum and subsequently decline steadily. Messenger M also continues to decline. It appears that the model-system at such a low level of external nutrient pool P_e is unable to maintain its functional organization and will decay. The problem will be discussed in more detail in a later section of the book.

It is of interest to follow the sequence of events further. Since P_i is the first reaction step from pool P_e (see equation 16, Table III), it is not surprising that the first concentration change takes place in pool P_i. In Fig. 26, the changes in pools P_n and P_a are indicated. It is evident that the changes are less drastic, but they follow closely the changes in P_i. This is to be expected, since P_a and P_n are formed by the next sequential reaction steps. The changes of free enzyme concentration reflect the changes occurring in the pools. It appears that external pool reduction is reflected in internal pools, and the growth of functional elements follows the pool concentration changes. This, of course, is to be expected, since the pools represent the basic elements by which the functional entities are built up. If we reduce the concentration of building blocks, it is obvious that there is reduced synthesis of functional entities. Sequential changes occurring in the concentration of various functional entities are too complex for detailed analysis. The concentration of an entity depends on the rate of new synthesis as well as on the decay. For example, Fig. 25 reveals that enzyme E concentration is reduced more rapidly than that of transport RNA C and ribosome B. This is to be expected, because, in the model-system, ribosomes and transport RNA are more stable entities than are enzymes.

f. Shift-Up Experiments

Shift-up experiments are limited by certain features of computer technology. Mainly, the rapid increase of external pool P_e causes a rapid growth of all functional elements, and these exceed the voltage limits (100 V) of the computer elements. For this reason, it was necessary to reduce rate constants k_{18}, k_{19}, and k_{20} and external pool P_e. After suitable reduction, it was possible to carry out experiments under conditions in which the external pool concentration would be doubled ($P_e = 2.0$). Figure 27 exhibits some of the functional entities after the increase of external pool value to $P_e = 1.5$, at the

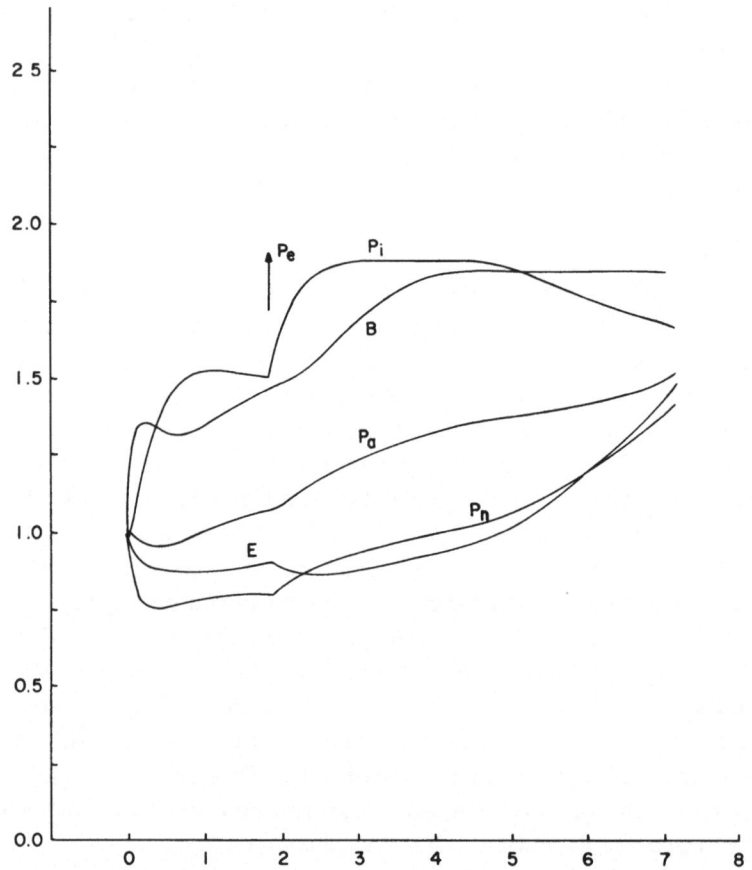

Fig. 27. Shift-up experiment. At the time indicated by arrow, pool P_e is increased 50%. The effect on: Pools P_i, P_a, P_n, ribosome B, and enzyme E.

time indicated by the arrow. It is evident that pool P_i increases rapidly, reaches a maximum, and then declines slowly. At the same time, pool P_n is also increasing continuously, but not at a uniform rate. Initially, there is an increase, then a decline, and at the end of the generation time there is again an increase in pool P_a formation. It is evident that the increase of ribosome B concentration is delayed when compared to the increase of P_i. In about half a generation time, B concentration reaches a constant value, and this level is maintained up to the end of a generation time. In order to analyze the effect of pool P_e increase more specifically, it was decided to increase the external pool to double value and to observe the behavior of all principal functional entities. Figures 28-30 describe this experiment. In general, growth characteristics are the same for Figs. 27 and 28, except that in the latter part of the growth phase ribosome B continues to increase, and there is a more drastic reduction of E after initial transient when the pool is increased. It is quite remarkable

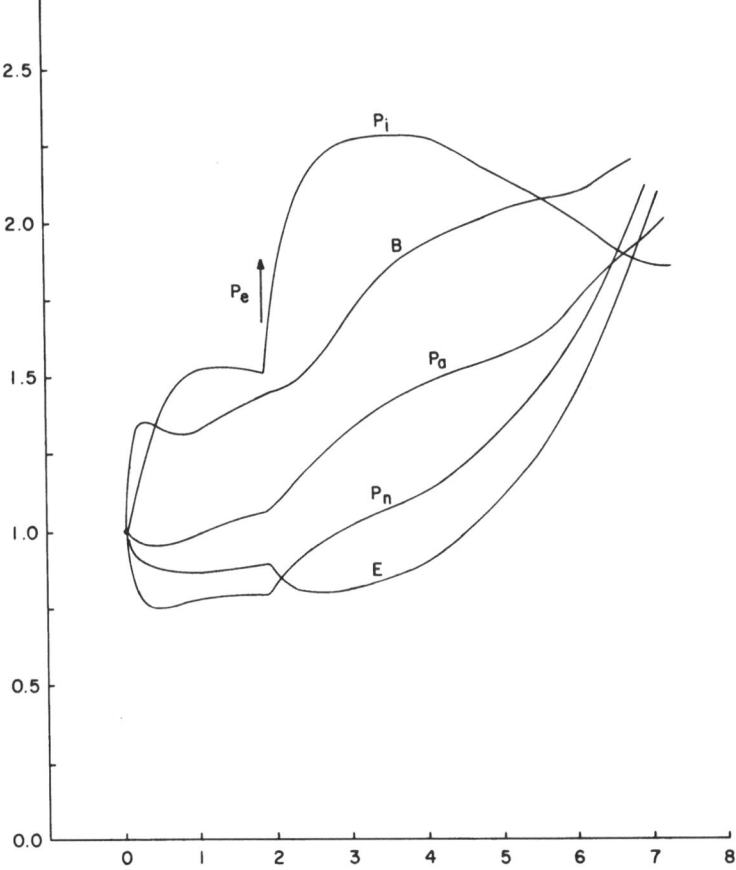

Fig. 28. Shift-up experiment. At the time indicated by arrow, pool P_e is increased 100%. The effect on: Pools P_i, P_a, P_n, enzyme E, and ribosome B.

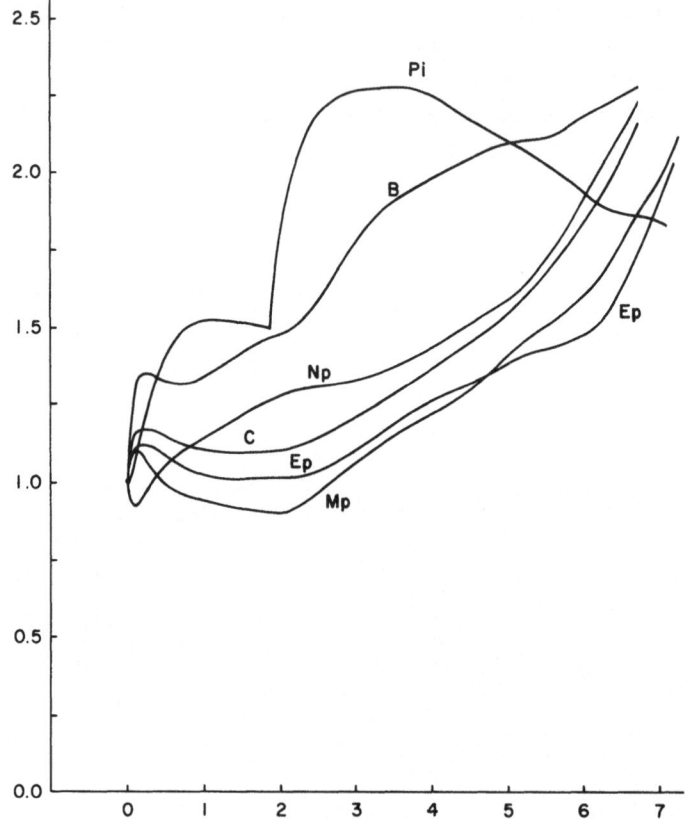

Fig. 29. Same shift-up experiment as in Fig. 28. The effect on: Messenger M_p, template N_p, enzyme E_p, transport RNA C, ribosome B, and pool P_i.

that pool P_i, after a twofold increase of pool P_e, still reaches the maximum value at a certain time and subsequently will decline. This seems to be a persistent characteristic of pool P_i in these experimental conditions. However, observations with an extended time scale show that this pool value passes through a minimum and then starts to increase again. Figure 30 shows that enzyme E is reduced simultaneously with the increase of the pool P_i concentration. This again results from the regulatory compensation of the complex [E P_i], and enzyme E concentration will subsequently increase again. It can be

observed that ribosome B and transport RNA C are increasing
only after some delay.   After further delay, messenger M
increases, and is followed by the increase of template N concentration.   Figure 29 shows the effect of pool P_e increase on
polymerase E_p synthesis.   Here, messenger M_p, template
N_p, and polymerase E_p all reveal an increasing growth trend
when pool P_e is increased. Polymerase E_p is not included in
a feedback inhibition system, and therefore there is no reduction
of E_p when the pool value is decreased. After the increase of
pool P_e all entities pass through a transient and then start to
grow, at first at a lower rate and subsequently at a faster rate.

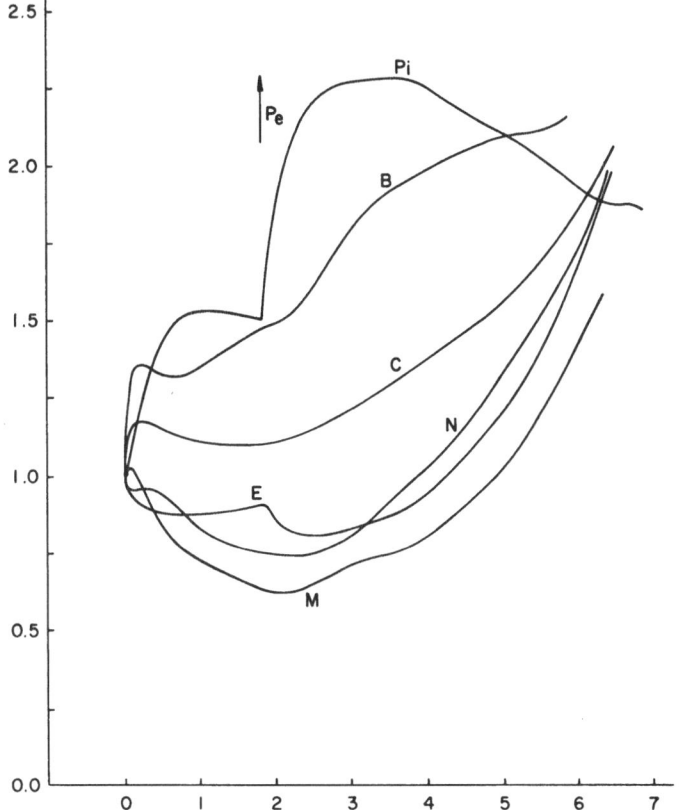

Fig. 30. Same shift-up experiment as in Fig. 28.   The effect on: Messenger M,
template N, enzyme E, transport RNA C, ribosome B, and pool P_i.

A systematic study of the effect of pool P_e concentration
on pool P_a is indicated in Fig. 31. Here, the external pool
concentration is either increased or decreased, and the normal
growth conditions are indicated by symbol "0" on the curve.
When P_e, which has a normal value of unity, is increased by
fractions of 0.25 and 0.5, then this reflects also in the increase
in P_a, as indicated by two upper curves (Fig.31). When pool
P_e is reduced by a fraction 0.25, then there is very slow growth

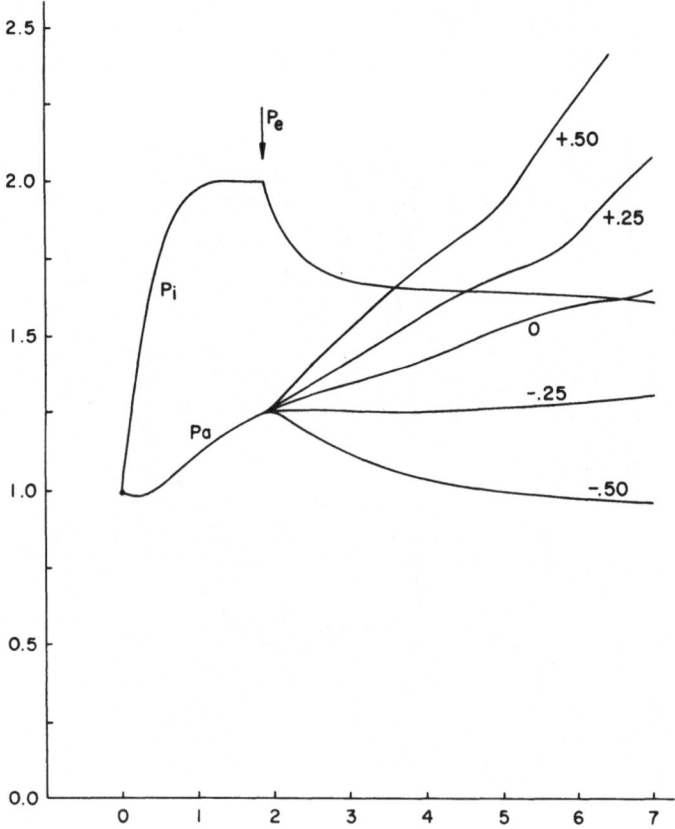

Fig. 31. Shift-down and shift-up experiment. Pool P_e reduced or increased and
pool P_a value is measured. Figures on the curve indicate the amount of decrease
(-) or increase (+) of P_e from value of unity. Normal P_a growth is indicated by
curve marked by zero. P_i is recorded when P_e is reduced 25%.

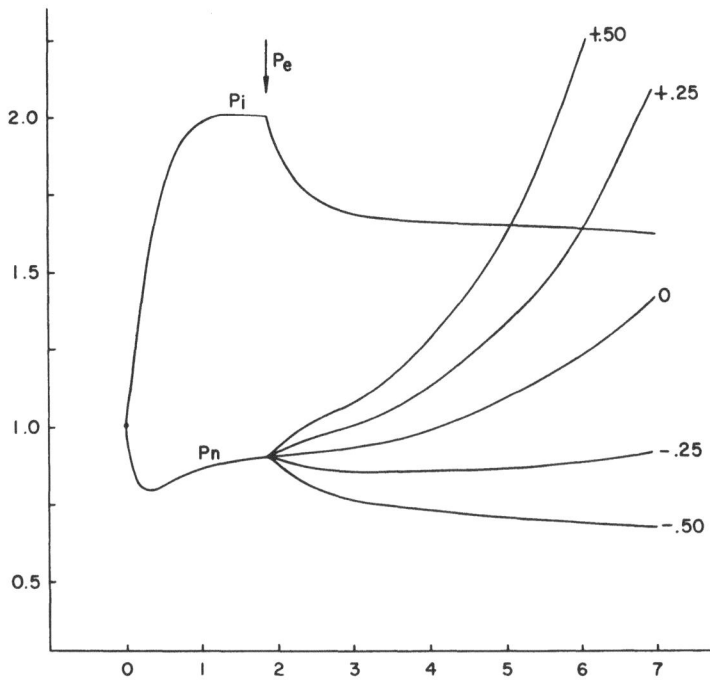

Fig. 32. Pool P_n recorded in the same conditions as in Fig. 31.

of P_a. However, at a 50% reduction of pool P_e (lowest curve on Fig. 31), there is actually a decline in P_a concentration. Ribosome B and pool P_n were recorded under similar conditions, for which data are presented in Figs. 32 and 33. While internal pool P_i follows the change in external pool P_e concentration closely, it is evident that there is always a lag phase between ribosome concentration change and the change of the pool value P_e. This is to be expected, because a certain amount of time is required for the synthesis of ribosomes. In summary, one can say that external pool changes up or down can be tolerated by the model-system in a certain range. However, when pool reduction is excessive, the system starts to decay. Similar phenomena can also be observed in actual cellular growth in a culture medium, where excessive reduction or increase of pool concentration will terminate the cell growth.

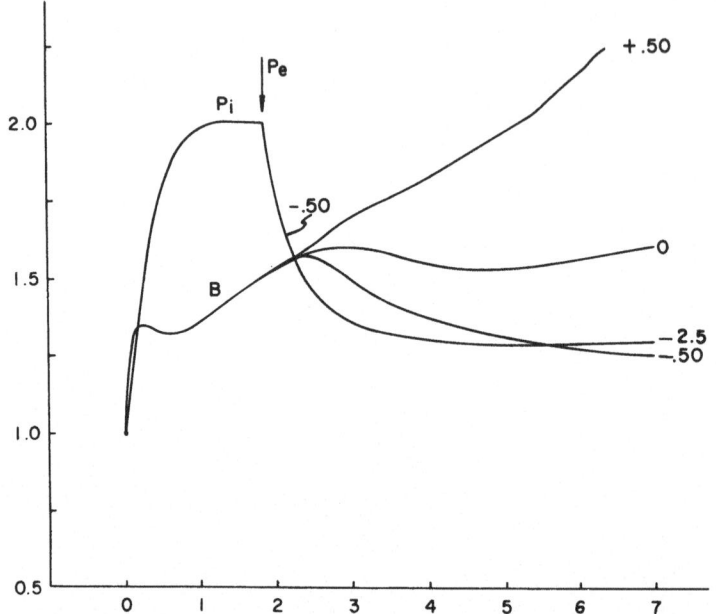

Fig. 33. P_e is reduced or increased 50% and ribosome B recorded. P_i is recorded when P_e is reduced 50%.

5. SIMULATION OF CELLULAR DORMANCY AND ACTIVATION OF GROWTH

a. Introduction

Normal cells in a proper nutritional environment usually grow and multiply. When cells are part of an organism, the rate of cell division varies in a wide range in various tissues. Some cells grow and divide rapidly, others slowly, and some do not multiply at all. Furthermore, some cells have lost their ability to multiply. These cells perform specific functions for a certain time and subsequently decompose. Cells acquire some particular kind of property for specific functions through the process of differentiation. Thus, specialized cell groups make it possible to develop highly organized structures and organisms. It is of interest to analyze the conditions which lead to cell growth and cell division, as well as the conditions which are required to terminate further growth. Cellular dormancy is a widespread phenomenon throughout nature. It can take various forms, for example, in spore formation where all potential growth of the cell is contained in a certain volume in an inactive state. Activation requires specific environmental conditions. Similarly, seeds contain all the genetic information and the capability to develop later into complex biological species, provided that the suitable environment is present. The dormant state can be maintained for a long time in conditions where growth is not possible. However, multicellular organisms usually contain cells which do not grow or divide in spite of the fact that they exist in a highly active metabolic environment. Such an extreme example is the nerve cell, which has a long lifetime and undergoes no multiplication. It is of particular interest that a complex organism can control synthesis differentially among various cells or limit synthesis throughout the organism for a period of time. Hibernation, where cellular division and general metabolic activities are reduced to a very low level by external temperature change, is one of many such phenomena. The primary site and the mode of dormancy control may not reside in the principal part of the system, but the effect of temperature change may operate on the special regulatory elements. However, in simpler biological species it is sufficient that temperature change takes place for the total system to go directly into a state of dormancy.

Evidently there is a wide range and degree of dormancy in nature, but all forms of dormancy operate via cellular activity. Here we shall attempt to analyze certain features of cell dormancy on the basis of a model-system. The principal issue seems to be to obtain a mode of control, so that it would be possible to achieve a reduction of cellular activity and the concentration of functional entities without disorganizing the system. In an advanced biological species, the regulation of cellular growth and division are highly complex processes. In a multicellular system, where multiple sets of highly interdependent cells operate in a general framework, it is essential that local as well as distant control mechanisms be in operation. First, there are two major types of controls: these are for cell division and for cell growth. In systems where cell division does not operate, let us say in the nerve cells, we have a condition of cellular dormancy throughout the life-cycle of the cell. Similarly, tissues which have a slow replacement rate have, in general, a strong growth control. We shall attempt here to analyze growth control at first on a conceptual basis and subsequently by quantitative studies. Division control will be considered later, but only on a descriptive basis. There are a multitude of regulatory mechanisms operating in the cells at various levels. Basic functional entities, DNA, RNA, carbohydrates, and protein, make up the bulk of the cell mass. The amount of growth-regulatory compounds present is trivial, since these, like hormones, can be highly active in extremely minute concentrations. Since regulatory compounds play an effective role in initiating and terminating growth processes, one may ask, where are the primary sites of regulatory operation as far as the major functional entities are concerned? It is expected that a uniform and smooth regulatory control could be exerted in such a way that major functional elements would be reduced or increased simultaneously. Otherwise one could expect a buildup of excessive concentrations of some functional entities in certain intracellular regions, and this could lead to geometric distortions of the system. This, in turn, could lead to the redistribution of metabolic products and to a change of the local concentration of substrates. This suggests that an intracellular disorganization could result from a process in which there is not a uniform reduction of principal cellular entities.

On the basis of the preceding considerations, we consider that cellular dormancy or the reduction of cellular activity can be studied on a model-system basis, since the model-system contains all of the basic functional entities which make up most of the cell mass. Here we have to assume that hypothetical regulatory compounds which are not directly generated in the model-system can operate on the functional entities of the system. We can ask at what level regulatory processes could operate during cellular dormancy. One can consider that the basic processes involved are the synthesis of enzymes and RNA. These can be analyzed on the basis of induction and repression mechanisms. Induction and repression can be analyzed in two principal areas, the genetic level and template level. On the genetic level, we have induction and repression which is associated with RNA synthesis, while on the template level, there is no requirement for RNA synthesis, and only protein synthesis occurs. While genetic induction is well documented [6, 7, 8], there is less information on induction and repression on the template level. However, there is evidence that hormonal and substrate control [9, 10] are definitely exercised on the operational level where RNA synthesis is not required and thus may play an important part in cell growth and differentiation.

b. Growth Regulation at Template Level

As indicated in Fig. 1, we have two principal templates, templates N and N_p, which participate in enzyme E and E_p polymerase synthesis. We shall first study the control of synthesis by individual templates separately, after which the activity of both will be controlled simultaneously. Template N_p is composed of messenger M_p and ribosome B. Normally, template N_p is active. In Table II, equation 24 shows that when N_p interacts with the substrate s_i, it forms N'_p, which is a functionally nonactive template. The compound s_i can be considered to be either a substrate or a hormone. It is further considered that in some phase in cell growth another hormone or substrate is introduced into the system which is capable of reactivating the inactive template N'_p. This is indicated by equation 25 in Table II, where N'_p complexes with s'_i, and thereby reconverts N'_p into normal N_p. Similar reaction mechanisms are indicated for templates N and N' in equations 26 and 27. The process of activation implies that the cell has been in the dormant state and that it is able to grow again when an activating substrate or hormone is introduced into the cellular system. Induction of cellular dormancy and activation of growth are processes which may last much longer than normal cell growth generation time. In order to carry out experiments on the computer for a reasonable observation time, the k_{30} value was increased considerably. This produced a more rapidly decaying model-system. In performing the experiments, the computer solution was synchronized with certain electronic switching circuits which permitted us to introduce s_i and s'_i at any time into the system. Usually, s_i was introduced into the system when functional entities were growing in a normal manner. After s_i was introduced, it was maintained in the system indefinitely or it was maintained there only for a certain time interval. In some experiments, when the first substrate was removed from the system, a second substrate as activator was simultaneously introduced into the system. In comparison with cellular growth generation time, the processes of activation and inactivation of the templates were considered to be rapid.

The first set of experiments is carried out under conditions in which templates N and N_p are inactivated simultaneously by substrate s_i and in which both inactive templates are subsequently reactivated when the substrate s'_i is introduced into the system. A series of experiments was carried out in which the effect of template alteration on the concentration of principal functional entities was studied. In order to evaluate growth kinetics in the modified system, the results are compared with the normal system. In Fig. 34, pools P_i, P_a, and P_n are represented where P_i^*, P_a^*, and P_n^* indicate the normal controls. At the time indicated by the arrow, s_i is switched in, and after a certain time interval s_i is switched out and s'_i switched in. It is evident that at the moment when the s_i is switched in, pool P_i begins to decline rapidly. Similarly pool P_n, after a short delay, starts to decline and continues to decline so long as inactivator s_i is present. On the other hand, pool P_a continues to grow while P_i and P_n are declining. There is a prolonged delay before P_a also starts to decline. Using an extended time scale, it can be observed that all three pools slowly reach a definite level while the activator s_i is maintained

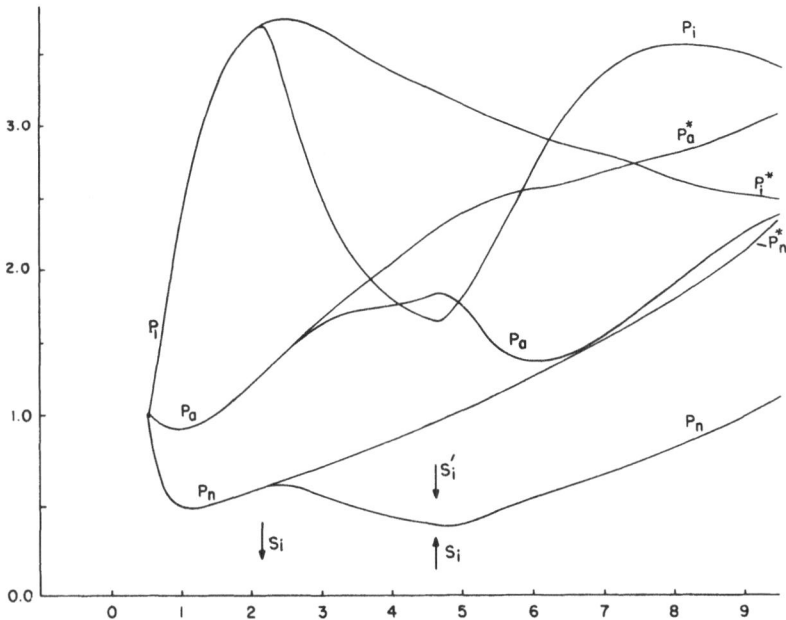

Fig. 34. Inactivation and reactivation of templates N and N_p. Normal components indicated with star. The effect on: Pools P_i, P_a, and P_n.

in the system. When substrate s_i is removed and activator s_i' is introduced in the system, the first result is the rapid increase of pool P_i, and, after a time lag, pool P_n starts also to increase. However, in contrast, pool P_a starts to decrease, reaches a minimum, and then starts to increase again. The behavior characteristics of pool P_a are complex, but the primary effect results from the rapid increase of transport RNA C, as indicated in Fig. 35, which causes the formation of $[C P_a]$ complex. Consequently, rapid reduction of P_a occurs. Figure 35 indicates that after the introduction of s_i, enzyme E and enzyme E_p both finally reach a rather constant concentration level. However, transport RNA C continues to decline, but observation at extended time scale reveals that C is also leveling off after a certain period of time. Since observation time on the computer is limited, one has to accelerate the experiment, so that s_i' was introduced into the system before this condition was

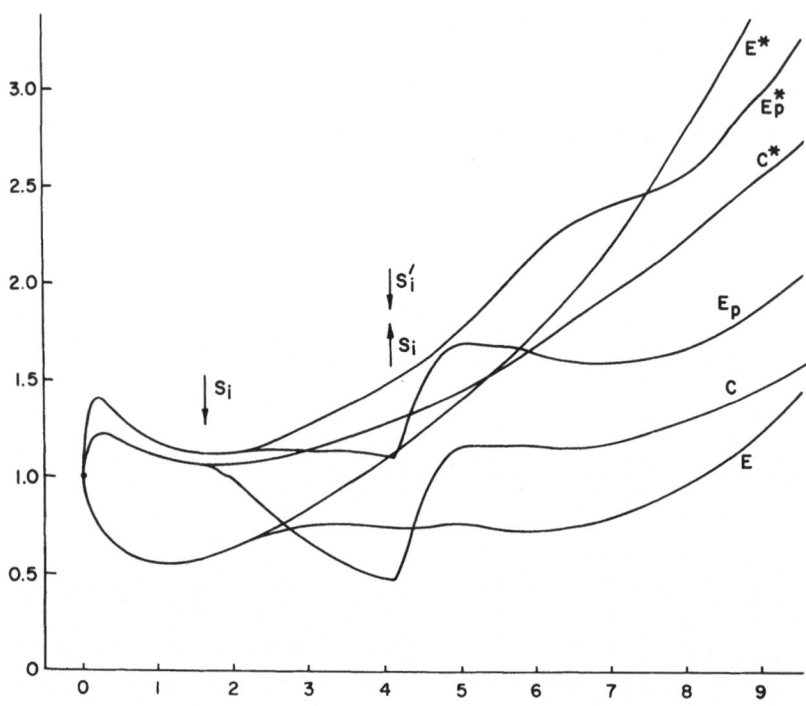

Fig. 35. Inactivation and reactivation of templates N and N_p. Normal components indicated by star. The effect on: Ribosome B, messenger M_p, and template N_p.

reached for C concentration. It is evident that after removal of s_i and introduction of activator s'_i rapid changes occur in the concentrations of functional entities. Pool P_i is increasing rapidly, and as a consequence, the buildup of regulatory complex $[P_iE]$ is also rapid. The concentration of enzyme E is maintained at a low level as long as P_i continues to rise. This indicates that the regulatory feedback loop operates very effectively. Figure 36 reveals the interrelationships between messenger M_p, template N_p, and ribosome B. It is evident that all these entities reach a constant level in the presence of s_i. The template N_p is reduced to an extremely low level, but nevertheless there is a definite amount of active template present in the system. Under these conditions, however, the concentrations of ribosome B and messenger M are maintained at high levels. The relationship between messenger M and template N under normal conditions and under the conditions of inhibition and reactivation are indicated in Fig. 37. When s_i is introduced into the system, there is rapid reduction of active template N until a low concentration level is reached. Messenger M is reduced at a lower rate but finally it establishes itself at a low concentration value. It is evident that the addition of the activator will produce a very rapid increase of messenger M and template N concentration. In order to follow the events in the time sequence, functional entities have been recorded in different combinations in Figs. 38 and 39. Figure 38 shows that after introduction of inhibitory substrate s_i, messenger M_p, template N_p, and polymerase E_p all reach, after a time interval, a rather constant concentration level. Since the template is the primary target of s_i, the concentration of N_p is

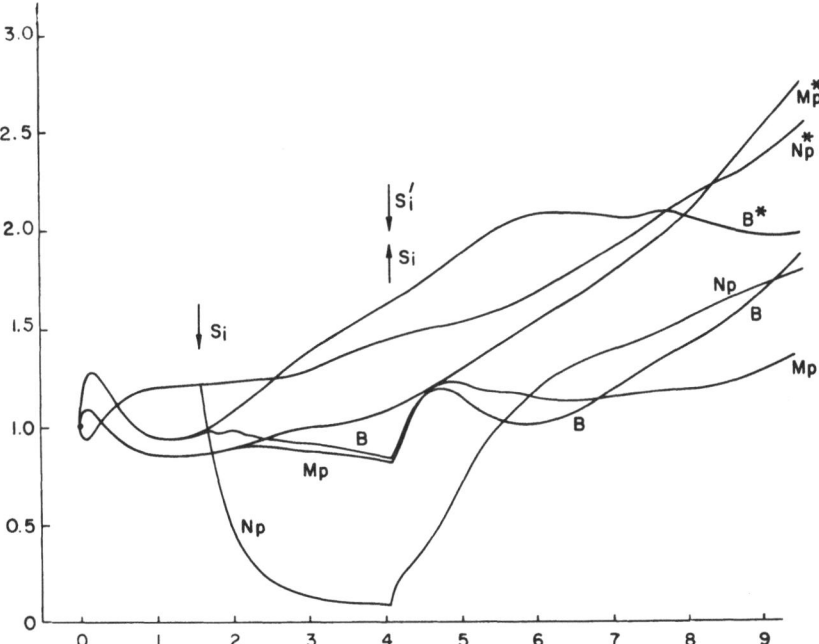

Fig. 36. Inactivation and reactivation of templates N and N_p. Normal components indicated by star. The effect on: Ribosome B, messenger M, and template N.

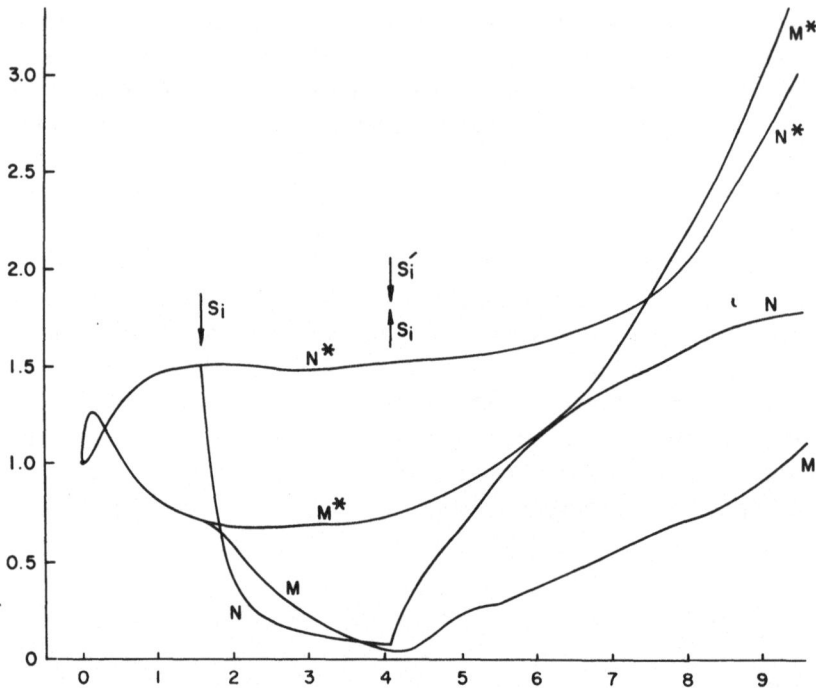

Fig. 37. Inactivation and reactivation of templates N and N_p. Normal components indicated by star. The effect on: Template N and messenger M.

reduced to a very low level. Since all these functional elements have a definite interrelationship, the changes occur in certain order when s_i' is introduced into the system. The sequence of events is as follows: at first, there is an increase of N_p, then M_p, and subsequently E_p. There is a considerable delay before P_{P_n} starts to increase. The interrelationships between ribosome B, transport RNA C, and the messenger M are indicated in Fig. 39.

These experiments reveal that the introduction of an inactivating substrate or hormone at template level is highly effective for controlling all-over cellular activity and growth. Furthermore, the concentration of various functional entities is reduced very smoothly and no disorganization occurs in the system. Similarly, the introduction of an activating substrate will permit uniform growth of all functional entities. Presented

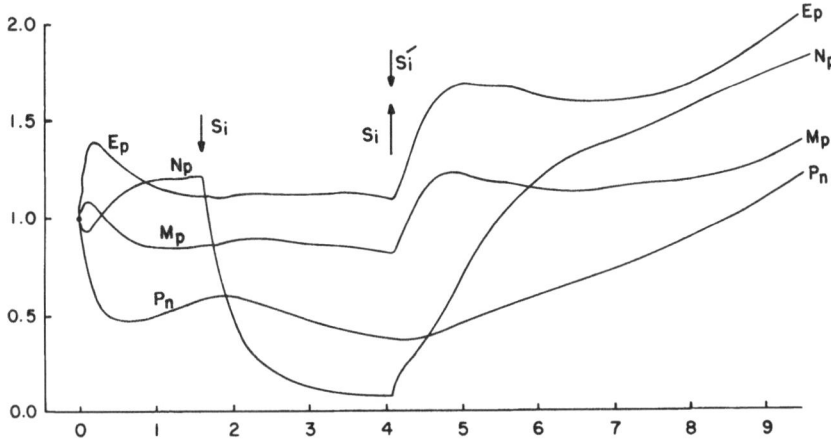

Fig. 38. Inactivation and reactivation of templates N and N_p. The effect on: E_p polymerase, pool P_n, template N_p, and messenger M_p.

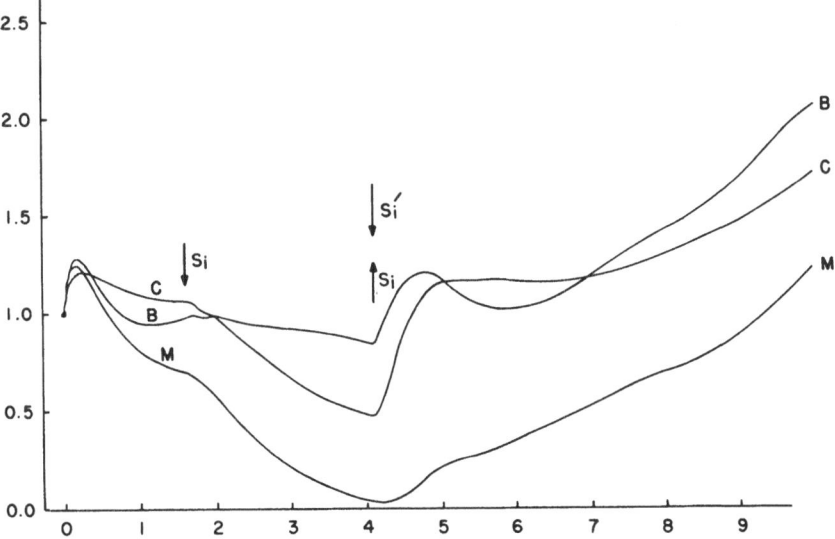

Fig. 39. Inactivation and reactivation of templates N and N_p. Interrelationships between messenger M, transport RNA C, and ribosome B.

experiments show that the model-system has indeed strong self-organizing characteristics, being capable of adjusting itself to rapid inactivation and activation of major functional entities. This further demonstrates that an enforcement-type control mechanism can be operative in both a positive and negative direction. Such controls are operative in cells where we have inducer and repressor type regulation. The principal effect of controlling template activity has a direct effect on protein synthesis. Furthermore, protein synthesis is a highly effective means of controlling all-over cellular growth. A similar conclusion was reached previously when we analyzed the effect of external pool level on protein formation. It is evident that when control is exercised on an enzyme synthesis level, it is possible to carry out a smooth regulation of growth by varying substrate

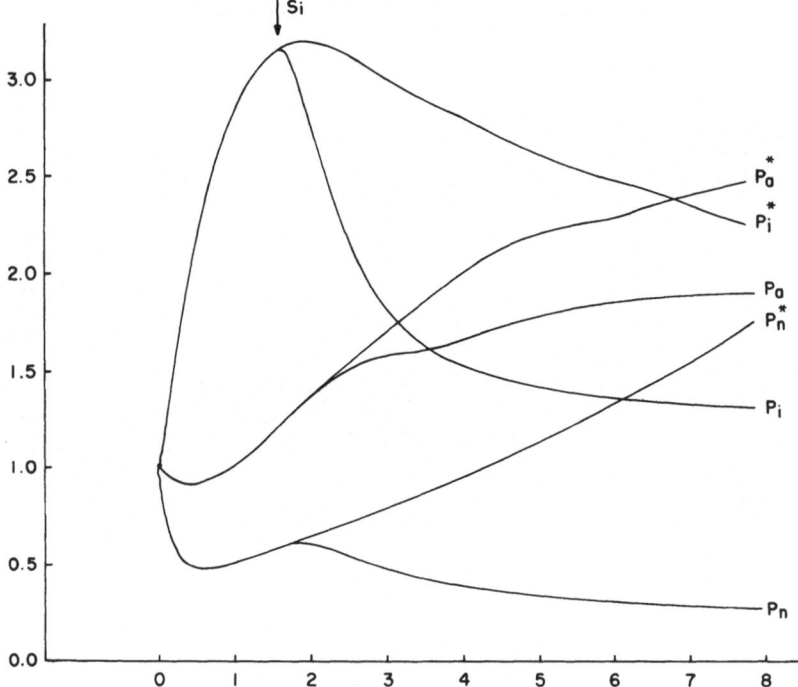

Fig. 40. Continuous repression of templates N and N_p. Pools P_i, P_a, and P_n in normal and repressed states. Stars indicate the normal state.

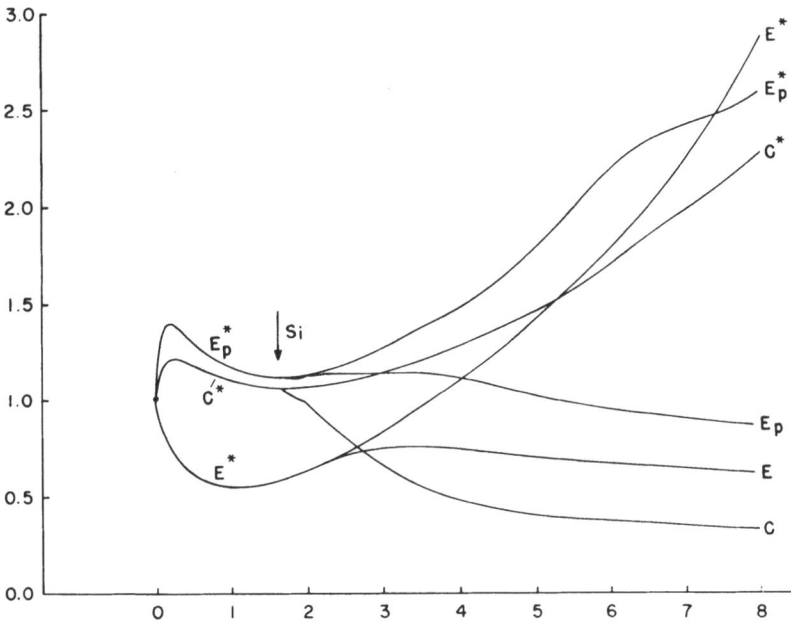

Fig. 41. Continuous repression of templates N and N_p. Enzymes E and E_p and transport RNA C in normal and repressed states. Stars indicate normal state.

s_i concentration in a wide range.   For example, it could be
reduced to a level of slow growth which may be sufficient for
cell replacement in tissues and organs. However, the system
could be activated for a rapid replacement.

Let us now consider the conditions under which growth of
the model-system is repressed for extended periods. These
experiments are carried out in the following manner: substrate
s_i is introduced into the system and maintained there without
subsequent activation.   Here, both templates N and N_p are
inactivated as in the previous experiment.   Figure 40 shows
the effect on the pool concentrations, where P_a^*, P_i^*, and P_n^* are
pools during normal growth. The concentration of s_i is selected
so that only growth repression occurs, but not complete growth
termination.   Pools P_i and P_n, after initial decline, finally reach
a constant level.   Pool P_a continues to rise until it establishes
itself at a definite level during the generation time.   Figure 41

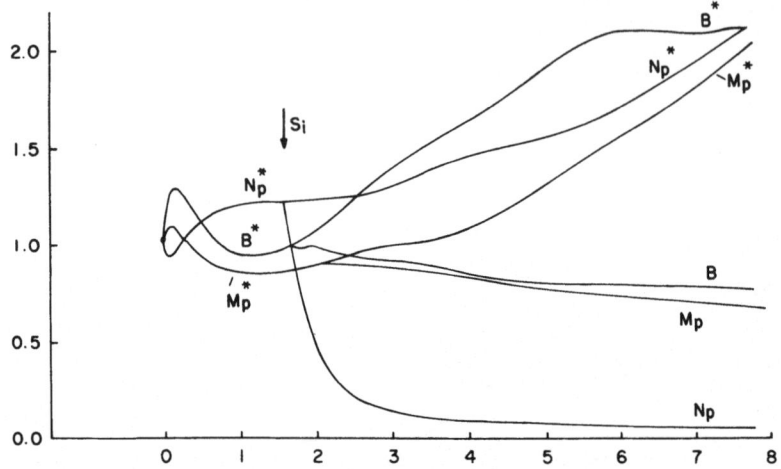

Fig. 42. Continuous repression of templates N and N_p. Messenger M_p, template N_p, and ribosome B in normal and repressed states. Stars indicate normal state.

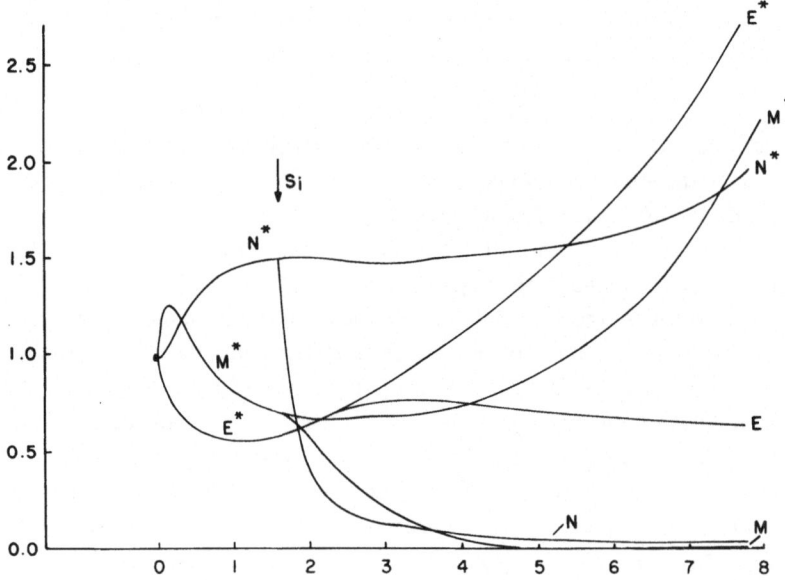

Fig. 43. Continuous repression of templates at N and N_p, enzyme E, messenger M, and template N in normal and repressed state. Stars indicate the normal state.

shows the kinetic behavior of enzymes and transport RNA. Again, symbols indicated by starts are the normal growth curves. At the introduction of the s_i, enzymes E_p and E reveal a small increase, which is followed by slow decline, and finally a relatively constant level for both is established. Transport RNA C is drastically reduced and establishes itself at a low level. Figure 42 shows the concentration or ribosome B, messenger M_p, and template N_p. It is indicated that ribosome B and messenger M_p concentration is maintained at a relatively constant level, while template N_p is maintained at an extremely low level. Figure 43 shows enzyme E concentration in relation to the messenger M and template N. It is evident that enzyme concentration is maintained at an inter-

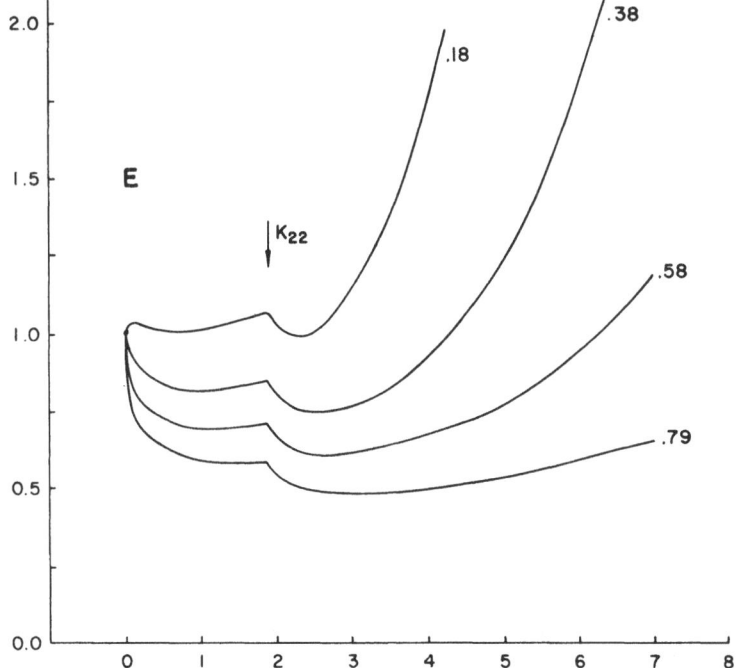

Fig. 44. The effect of rate constant k_{22} on enzyme E concentration. After certain periods of growth, rate constant k_{18} is increased to double value, which is equivalent to doubling of external pool, and enzyme concentration is studied at various values of k_{22} (equation 22, Table III).

mediate level, while messenger and template levels are very low. In summary, it appears that cellular growth can be very effectively controlled on the template level and growth repression can be quite extensive, so that drastic reductions occur in concentrations of functional entities. This kind condition can be maintained indefinitely as observations on computer with extended time scale indicate. While this analysis shows that the repressor molecule at the template level is effective in accomplishing a thorough control action, the source of the repressor molecule is not indicated here. For a more complex model-system it can be visualized to be part of a regulatory circuit involving additional genes.

c. Growth Control on the Level of Regulatory Complex $[EP_i]$

It was of interest to consider some other features of control mechanism in the system, namely, the regulation which arises from enzyme activity. The feedback loop between enzyme E and pool P_i (Table II, equation 22) has such operational characteristics and serves as a control mechanism.

It is widely recognized that regulatory features are essential for the orderly growth of a cell, especially when the cell is in a condition in which it is exposed to large variations of external factors or external pools. For this purpose, experiments were carried out to observe enzyme growth under conditions in which regulatory interaction on k_{22} level was varied in a wide range. In order to explore the effectiveness of the regulatory mechanism, the system was tested under rather severe experimental conditions. It was of interest to see how this control mechanism operates when the value of the external pool P_e is suddenly doubled under conditions where k_{-22} is kept constant while k_{22} is varied. Figure 44 shows the experiment in which external pool P_e, at a certain time in growth, was suddenly doubled in value, the effect of various k_{22} values are subsequently studied. Numbers on the curve represent the relative k_{22} values. After the introduction of the increased pool P_e, as indicated by the arrow, the growth of the enzyme varies in a wide range, depending on k_{22} values. It is indicated that k_{22} very effectively controls the rate of free enzyme concentration, and that it affects the growth rate as well. When k_{22} values are increased, the growth rate is largely decreased. In contrast, at the low values there is a rapid increase of enzyme formation as well as a more marked initial transient. This transient results from the complex [$E P_i$], when P_i suddenly increases. It is evident that the feedback mechanism, where interaction takes place between the enzyme and substrate, is very effective in controlling the growth rate. In order to study this mechanism further, the external pool concentration was suddenly varied in

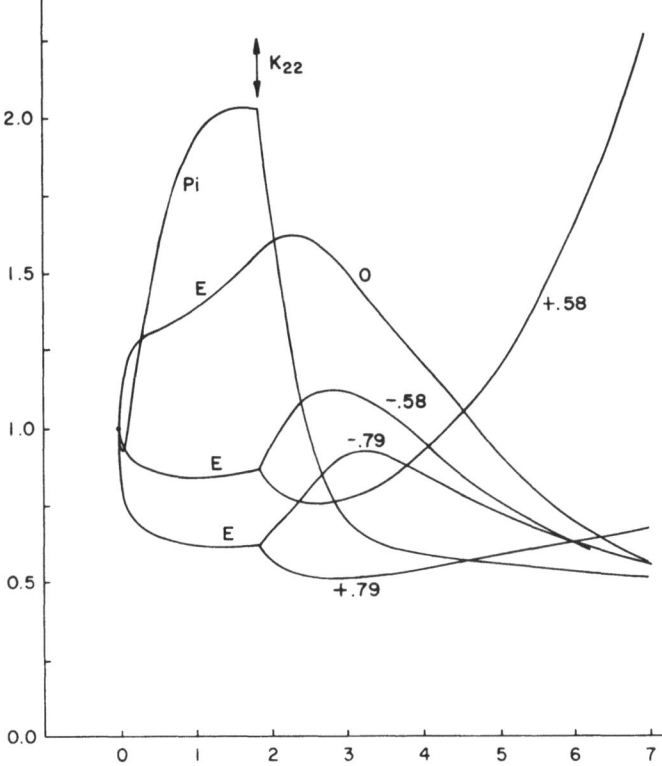

Fig. 45. The effect of the variation ($^+_-$) of external pool P_e on enzyme E growth at various values of k_{22}. At time indicated by arrow, pool P_e is increased or decreased by 80%. Numbers on the curves show the values of k_{22} at the time of shift-up (+) and shift-down (-). Zero on the curve indicates that there is no feedback loop ($k_{22}=0$).

a positive or negative direction. Figure 45 shows such an experiment where P_e is varied ± 80%. Three different values were assigned to k_{22} (0, 0.58, 0.79). Pool P_i is indicated in Fig. 45 when the external pool was reduced at the time indicated by the arrow. It is indicated that if $k_{22} = 0$, enzyme E, which starts to grow from initial condition, still continues to grow for a while, passes through a maximum, and then starts to decline steadily. When P_e is increased, there is a very rapid growth of E, which is too large to be recorded. When there is a relatively strong feedback loop ($k_{22} = 0.58$), there is initially a rather rapid increase of enzyme concentration. However, the maximum reached, a steady decline takes place. When pool P_e suddenly increases, there is initially a reduction of free enzyme E. This is a regulatory effect of complex $[P_iE]$. After passing through a minimum, a rapid growth of the enzyme is indicated. In order to show the effect of excessive feedback control, we increase k_{22} to the value 0.79. This reduces the growth rate of enzyme E very drastically indeed. It is further evident that 80% reduction of pool P_e is too extensive, since the system becomes nonviable at all three k_{22} values. It is evident that the regulatory mechanism operates very rapidly when sudden changes occur in external pool concentrations, although is cannot compensate for a long-term growth, when pool reduction is too extensive. On the other hand, in the range in which the control exists, it is very effective, slowing down the growth of enzymes when the external pool suddenly increases. If the regulatory mechanism is too strong, the system will be able to grow only very slowly, even in very suitable environments. These simulation experiments suggest that a cell which grows in an environment which varies widely can damp out by its own regulatory mechanisms rapid external variations. A strong regulatory mechanism is required in order for the system to grow smoothly in an environment which varies drastically. In the reverse case, systems which need rapid adjustment require that the regulatory mechanisms be relatively light. It appears that properly adjusted feedback loops serve this purpose very effectively.

d. Regulatory Processes in Dormancy

We have observed that an adequate growth control in a positive and negative direction can be exercised at the template level when both templates N and N_p are the subject of that control. The question arises as to whether it is possible to exercise that control on a more limited basis. While the products of templates N and N_p are both enzymes E and E_p respectively, the functions of both entities are distinctly different from the operational point of view. The function of E_p is to take part of the syntheses of messenger M which is a complementary part of template N. One can question whether E_p alone could control the cell growth process. Such a mechanism could be effective, economical, and direct, since E_p is a special type of enzyme, occupying a highly strategic position in the synthesis. In order to elucidate this point, a series of studies was carried out on the computer. The model-system was used as usual, except that the rate constants k_{26} and k_{27} were abolished. Therefore, template N operated normally. Template N_p was converted to N_p' by k_{24} and reactivated by k_{25}. In order to observe changes during one generation time, growth condi-

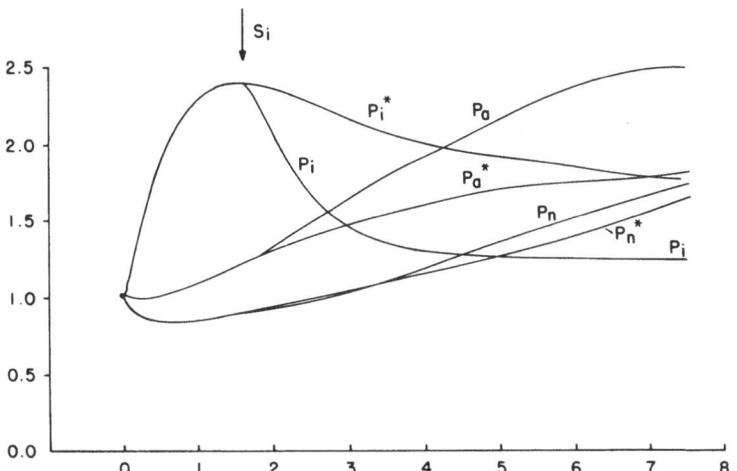

Fig. 46a. Continuous repression of template N_p. Pools P_i, P_a, and P_n in normal and repressed state. Stars indicate normal state.

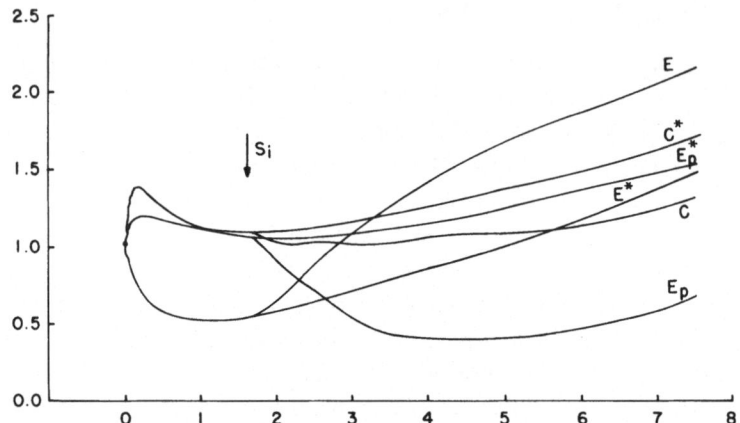

Fig. 46b. Continuous repression of template N_p. Enzymes E and E_p and transport
RNA C in normal and repressed state. Stars indicate normal state.

tions were adjusted to a slow rate. Figure 46a shows the interrelationship between the pools when at a certain time s_i is
introduced into the system. It is evident that after the
introduction of the inhibitor s_i, pool P_i is drastically reduced
and finally reaches a constant value. On the other hand, pool
P_a increases significantly above the normal value, and there

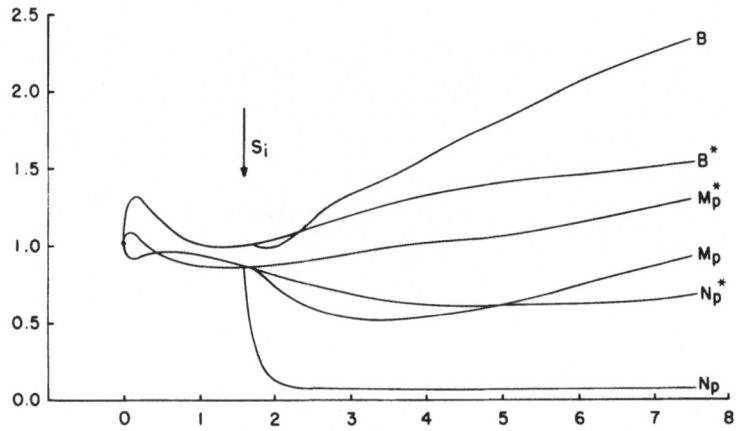

Fig. 46c. Continuous repression template N_p. Ribosome B, template N_p, and
messenger M_p in normal and repressed state. Stars indicate normal state.

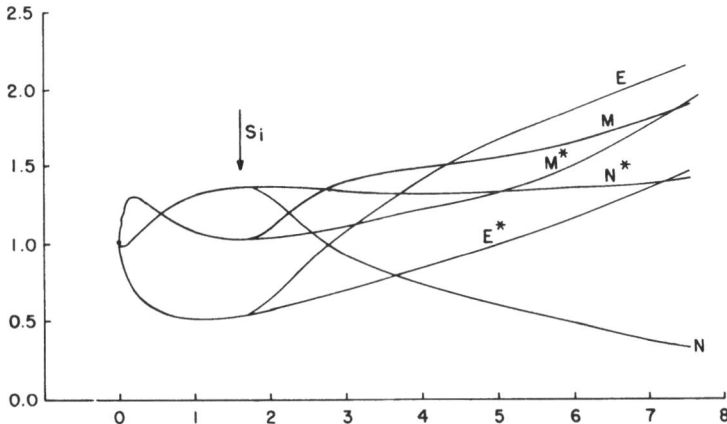

Fig. 46d. Continuous repression of template N_p. Enzyme E, template N, and messenger M in normal and repressed states. Stars indicate normal state.

is also a slight increase in pool P_n concentration. It appears that pool adjustment after the introduction of the inhibitor is not uniform. Figure 46b shows that enzyme E concentration after the introduction of the inhibitor is remarkably increased above the normal value. E_p polymerase is dramatically reduced, and after reaching a minimum value it starts slowly to increase. The growth of transport RNA C is to some degree reduced in the presence of the inhibitor. Figure 46c reveals the relationships between template N_p, messenger M_p, and ribosome B. It is evident that ribosome B concentration increases drastically when inhibitor s_i is introduced, while template N_p concentration is highly reduced. There is also a definite reduction of messenger M_p. Figure 46d shows the interrelationship between N, M, and E. It is evident that while template N is reduced, both messenger M and enzyme E increase above the normal level. The principal cause for the increased synthesis of M and E is the increase of pools P_n and P_a (Fig. 46a). The all-over picture resulting from this experiment is that there is no uniform control by the inhibition of the growth process at N_p level alone. The system exhibits a rather disorganized state where some components go up, while the others are reduced. It is of interest to follow the

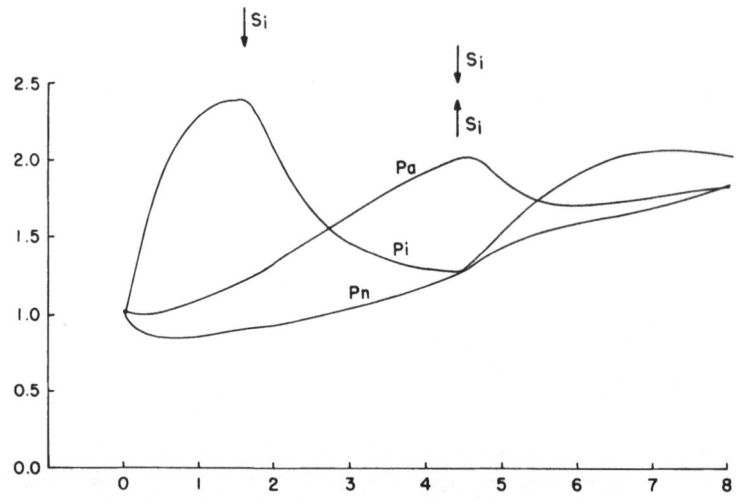

Fig. 47. Repression and derepression at template N_p level. The effect on: Pools P_a, P_i, and P_n.

results when inactive template N_p' is converted back to N_p via reaction step k_{25}, Table II. The first experiment is shown in Figs. 47–50. Figure 47 shows that when s_i' is introduced into the system, there is initially a decrease of pool P_a, but it gradually starts to increase again. There is a steady increase of P_i which finally reaches a relatively constant level. Pool P_n grows rapidly initially, but later continues to increase at a lower rate. Figure 48 shows that activation of N_p initially causes a reduction of enzyme E concentration, but after passing through a minimum, E will increase again. At the same time, there is initially a drastic increase in enzyme E_p and transport RNA C concentration, but at the later phase, the growth rate is reduced. Figure 49 shows the messenger M_p, template N_p, and ribosome B concentration. It is evident that introduction of s_i' will produce a transitory increase in ribosome concentration, which, after a decline, starts again to increase. There is a rapid intial increase of template N_p and messenger M_p, followed by slow growth. These entities have growth characteristics similar to E_p (Fig. 48). Figure 50

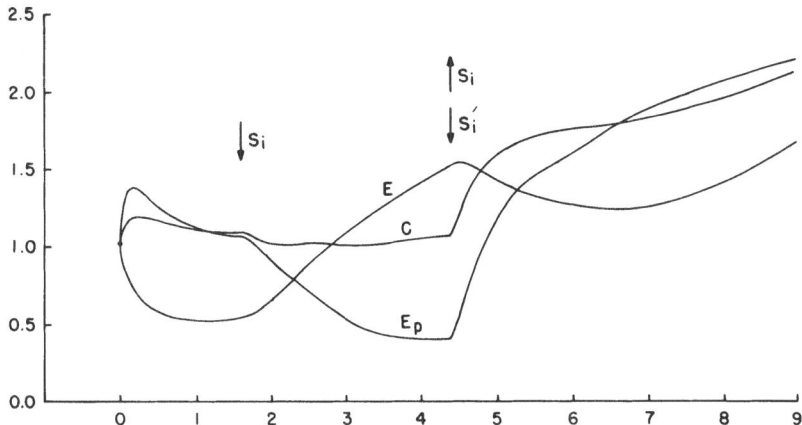

Fig. 48. Repression and derepression at template N_p level. The effect on: Enzymes E and E_p and transport RNA C.

shows that s_i' introduction initially produces a prolonged reduction of enzyme E concentration, but later E starts to increase. Both N and M, after a small initial decline, increase steadily. The overall effect appears to be again a nonuniform growth response after the introduction of s_i'. In summary, one could say that the application of control at the polymerase

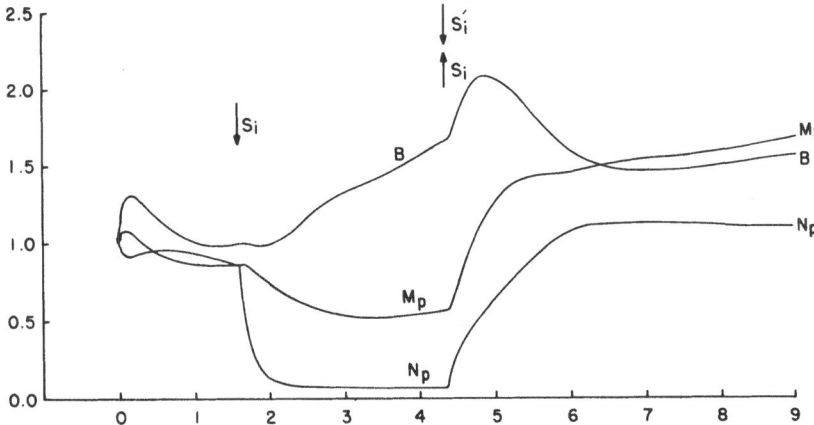

Fig. 49. Repression and derepression at template N_p level. The effect on: Messenger M_p, template N_p, and ribosome B.

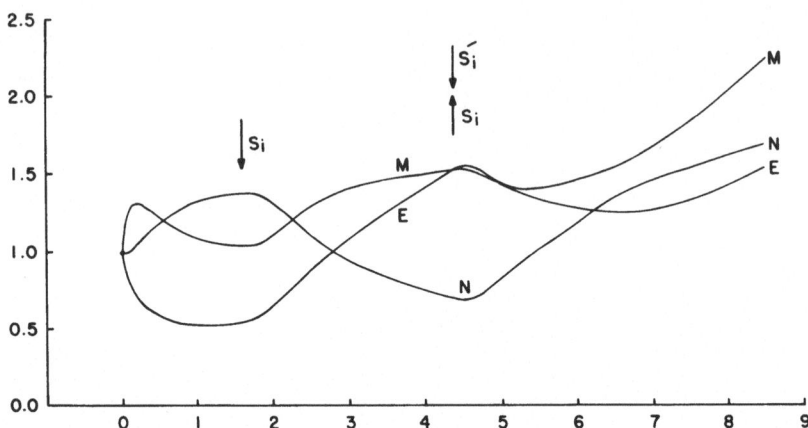

Fig. 50. Repression and derepression at template N_p level. The effect on: Messenger M, template N, and enzyme E.

template N_p level produces in the model-system very drastic changes in concentrations of functional entities.

Since it was pointed out previously that drastic changes are not desirable in growth regulation, one can conclude that an effective and smooth control is not possible on the template N_p level. The question arises as to whether effective growth control can be obtained on the template N level. A second experiment was carried out under condition in which the template N_p was normally operational, while control was exercised on the template N level by the application of substrates s_i via rate constant k_{26}. Preliminary experiments revealed that it is possible to exercise rather smooth control on the template N level alone. Therefore, a series of experiments was performed at different levels of template inhibition. It is of great significance that a smooth regulatory performance is exhibited even at levels where cellular growth is reduced to such a low level that the cell is gradually decaying, but rather in a smooth fashion. Such data are represented in Figs. 51–54. Figure 51 shows that all pool levels, after initial transient produced by the introduction of s_i into template N system, exhibit a decline. The course of the decline is rather uniform for all three pools, as long as the inhibitor continues to be

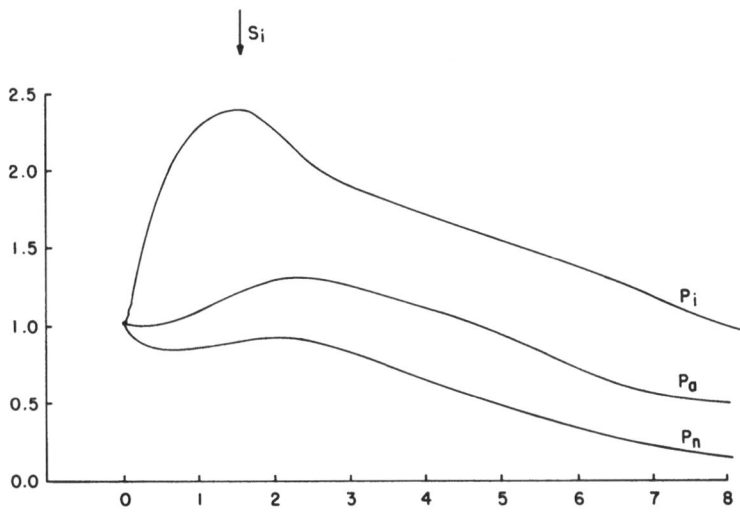

Fig. 51. Continuous repression of template N. The effect on: Pools P_i, P_a, and P_n.

present in the system.   Figure 52 shows that after initial transient, enzymes E_p, E, and transport RNA C all are declining in a rather smooth fashion.   Figure 53 shows that after initial increase, M_p and B continue to decline, while N_p decline is more rapid initially and then slowly approaches the zero concentration.   In a similar fashion the introduction of s_i rapidly reduces active template N concentration (Fig. 54).

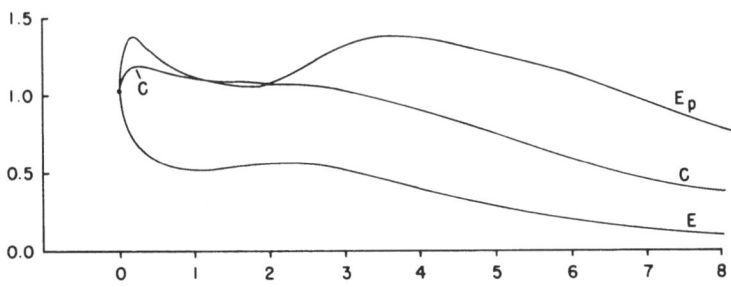

Fig. 52. Continuous repression of template N.   The effect on: Enzymes E and E_p and transport RNA C.

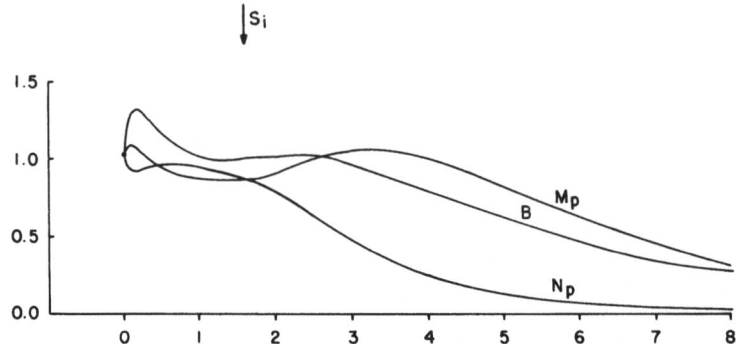

Fig. 53. Continuous repression of template N. The effect on: Messenger M, template N_p and ribosome B.

After a short delay, messenger M and enzyme E are also reduced to a very low level. It is of interest that all three components asymptotically approach each other near the zero level. Experiments reveal that the total enzyme synthesis, excluding the polymerase E_p, is capable of effectively controlling cell growth. Furthermore, the inhibition introduced at template N level into the model-system is capable of reducing synthesis to very low level without producing excessive gradients in the concentrations of functional entities. When observations are extended for several generation times, it is seen that the system finally ceases to exist. This of course means that the system has lost its original self-organizing

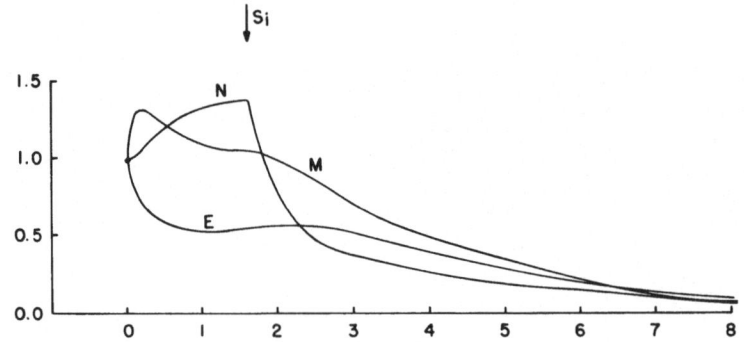

Fig. 54. Continuous repression of template N. The effect on: Messenger M, template N, and enzyme E.

characteristics or in terms of biological concepts "the cell has lost its viability." This subject will be analyzed in greater detailed in a subsequent section.

6. SELECTIVE INHIBITION AND ACTIVATION OF CELLULAR GROWTH

a. General Comments

The purpose of this study is to gain insight into the mode of action of various growth activators (such as hormones and sub-strates) or inhibitors (such as antibiotics and other drugs) in a living cell. For this purpose, various functional sites within the model-system are activated or inhibited, and the all-over effect observed in the model-system. It is considered that a particular agent can interact with any specific functional entity or interfere with the overall process of synthesis. Since this model-system was designed for a broad, general study of cell growth, it is possible to carry out interaction studies on the group property level. Direct effects, resulting from interactions on the level of certain functional entities, can be studied, but it is not possible to analyze certain compensatory phenomena, such as adaptation of the system to a new condition, by activation of some of the repressed genes. This type of phenomena can be studied, but the model-system has to be designed for this particular type of problem. Nevertheless, the present system permits us to simulate many biological experiments. For example, "leakiness" of cellular membranes can be simulated by the rate constant k_{30}. By changing rate constant values, the disappearance rate of internal pool P_i can be varied.

The action of various metabolites, antimetabolites, and antibiotics on microorganisms has been reviewed by Davis and Feingold [11]. It is evident that the site of actions of various agents is known in some cases rather specifically and in others more generally. For example, some antibiotics act on cell wall formation, while others affect the membranes; some affect protein synthesis, others, nucleic acid synthesis and metabolism, etc. No attempt will be made here to compare cellular studies directly with the experiments on the model-system. However, we hope that an analysis of the model-system will help to elucidate inhibitory and activating mechanisms in cellular systems.

b. The Effects at the Pool Level

The model-system indicates that there is a competition between various functional entities; for example, all four genes require P_n for RNA synthesis. There is also a competition between the pools P_n and P_a for the general internal pool P_i. We shall perform experiments to study the effect of factors which increase and decrease pool values on the overall behavior of the model-system.

Conversion of pool P_i into P_n requires the presence of enzyme system E_n, where the reactivity is represented by all-over rate constant k'_{19}. Since k_n determines the fraction of total enzyme E which exists as E_n, then total nucleotide pool P_n conversion is represented by a combined rate constant $k_{19} = k_n \cdot k'_{19}$ (Table II). Consequently, one can consider here that that pool variation can be caused by partial activation or inactivation of enzyme system E_n, in which case k_n would be changed. On the other hand, the specific transport property of the nuclear membrane could be reflected in constant k'_{19}. For example, variation of k'_{19} in the model-system could simulate the cellular conditions in which the decrease or increase of the transport properties of the nuclear membrane are affected by a hormone. In view of these considerations, experiments are carried out to explore the effect of the transient variation of P_n on the other entities in the model-system. Figure 55 shows the effect of pool P_n variation on the template N_p. At the time indicated by the arrow, the rate constant k_{19} is reduced by 50%. It is evident that pool P_n concentration is immediately reduced. After reaching a minimum value, P_n starts to grow again. In these experiments pool P_n concentration was varied over a wide range. There was a maximum increase of 30% and a maximum decrease of 90% from normal pool level. The figures on the curve N_p indicate the relative k_{19} values at various experiments. Curve "1.0" indicates the normal growth of template N_p. A 30% increase (curve "1.3") of k_{19} produces a significant increase in

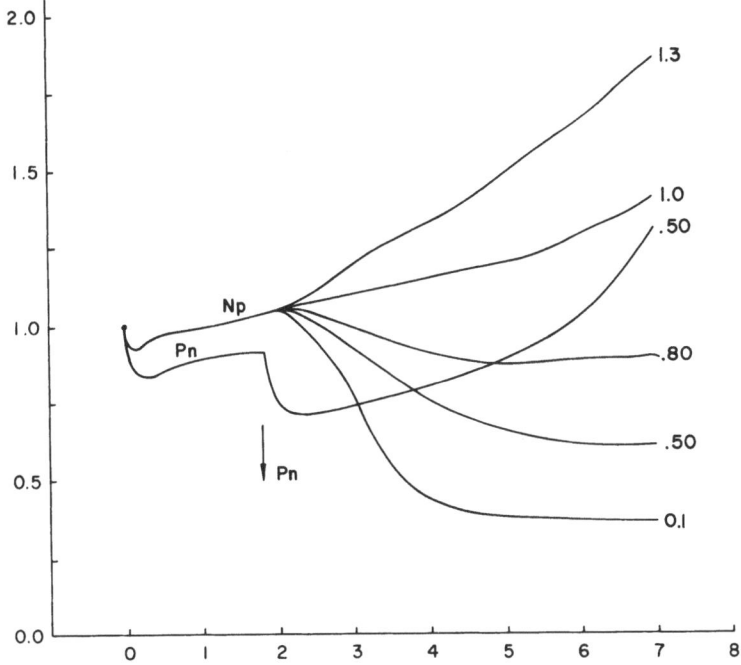

Fig. 55. The effect of pool P_n variation on template N_p formation. P_n reduced or increased at time indicated by arrow, by varying rate constant k_{19}. Figures on the curves show the changes in relative k_{19} values in relation to normal pool value ($k_{19} = 1.0$).

N_p growth. However, reduction of k_{19} by 50% practically arrests the N_p growth, while a 90% reduction of k_{19} not only produces drastic initial decline, but a slow reduction in N_p concentration follows as well. The effect of the N_p decline is reflected in Fig. 56, where the enzyme E_p concentration, at the same k_{19} value after the initial increase, suffers a reversal and continues to decrease rapidly (curve ".10"). Figure 56 indicates, in contrast, that a reduction of k_{19} by 50% is associated with an increase of E_p (curve ".50"). Further reduction of k_{19} by 75% and 90% reduce P_n growth extensively (curves ".25" and ".10"). However, it is really not obvious why E_p (at the value $k_{19} = .10$) will increase initially and then decline. It appears that functional organization may suffer considerable difficulty at the time pool

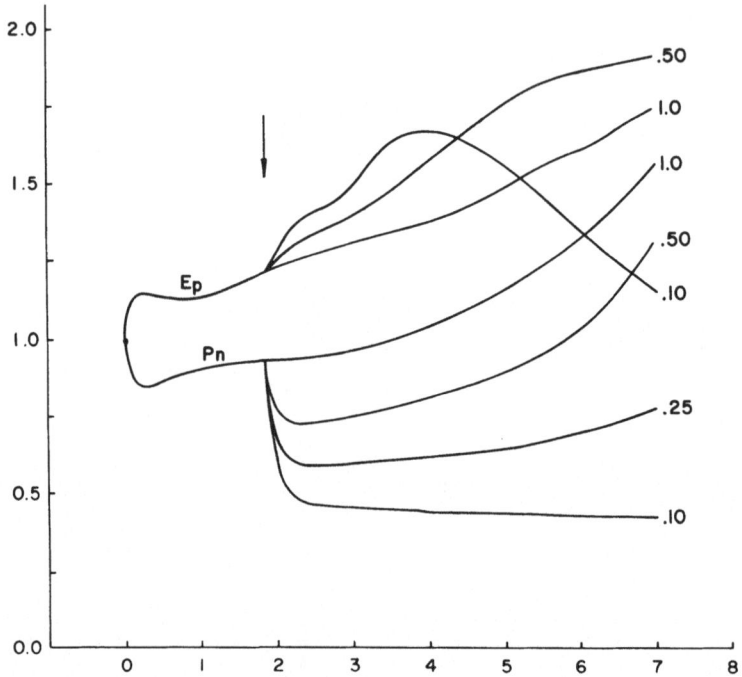

Fig. 56. The effect of pool P_n on E_p polymerase and P_n concentration. Conditions are the same as in Fig. 55.

P_n starts to decline. Figure 57 shows the same effects on enzyme E. The effects of a 50% reduction of k_{19} on various functional entities are presented in Fig. 58. There is an initial decline of ribosome B, enzyme E, and transport RNA C, but after passing through a minimum all these entities start to grow again. However, in contrast, pool P_a and enzyme E_p increase significantly at the same moment that k_{19} is reduced.

The effect of a k_{19} increase (25%) on several functional entities is shown in Fig. 59. It is evident that pool P_i concentration suffers a marked reduction initially, followed by slow decline. In contrast, pool P_n has an immediate rise and continues to increase steadily. Enzyme E, transport RNA C, and ribsome B all have a slight initial rise, followed by steady growth. Ribosome B concentration seems to level off at the end of generation time. Polymerase E_p, however, reveals an initial reduction in growth rate and subsequently continues to grow rather slowly.   This experiment indicates that a k_{19} increase produces alterations in the internal equilibrium of functional entities in the model-system, and transient effects have not disappeared during one generation time. It is evident that there is an initial increase in the concentration of most functional entities, but at what future time and at what level a balanced growth is established is not clear from these experiments.   Extensive and prolonged growth cannot be observed on the computer since concentrations of entities become too large.

A limited number of observations were carried out on growth in conditions where pool P_a was rapidly changed by varying k_{20}.   Formation of pool P_a, which is required for the protein synthesis, is expressed in Table III, equation 13. The term for pool synthesis is $k_{20}\ P_iE$, while the negative term, k_9CP_a, represents pool utilization. The term, k_9CP_a, represents the complex $[CP_a]$ formation and in our experiment is considered to be constant, in the sense that the k_9 value is not varied.   The variations in k_{20} in the model-system can be

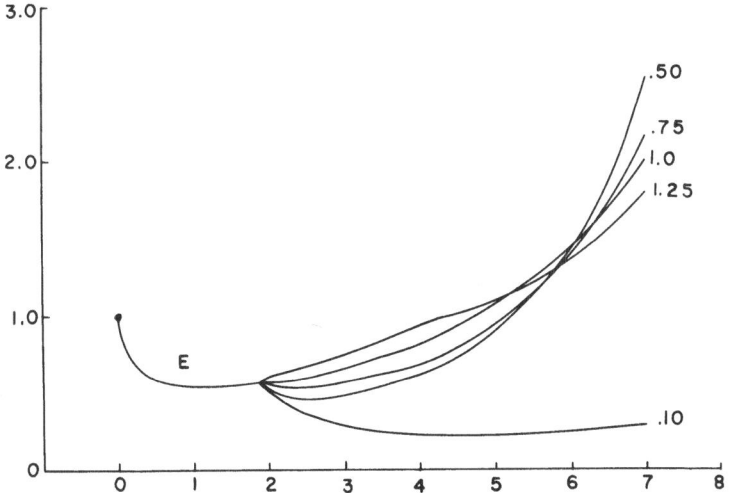

Fig. 57. The effect of pool P_n variation on enzyme E concentration. Conditions are the same as in Fig. 55.

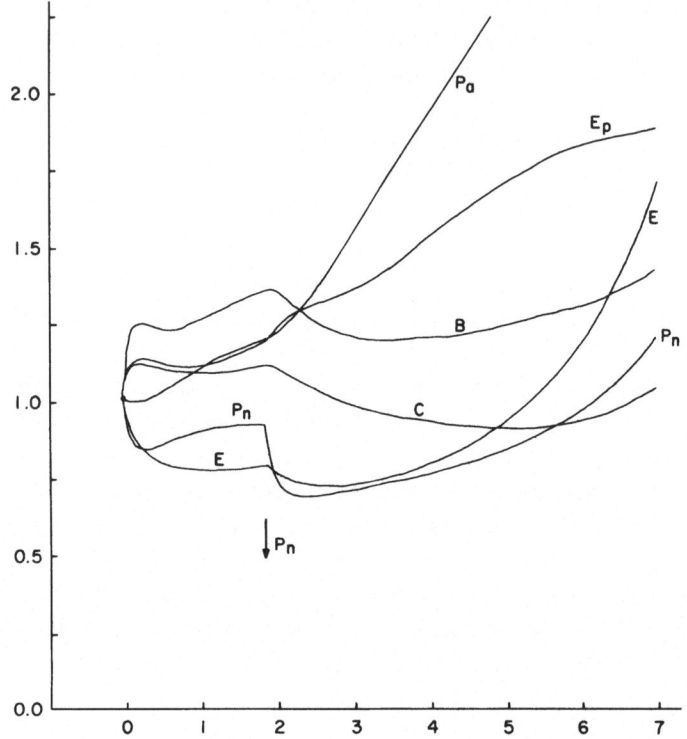

Fig. 58. The effect of 50% reduction of P_n concentration (at time indicated by arrow)
on: Pools P_a and P_n, enzymes E_p and E, ribosome B, and transport RNA C.

considered to be equivalent to activation or deactivation of an
enzyme in a cellular system, for example, by a substrate or an
external compound. First, we shall study the interrelationship
between pools P_a and P_n. Figure 60 shows that in normal state,
pool P_a grows steadily (curve "1.0"). When k_{20} is suddenly
increased by 25%, at the time indicated by the arrow, it is
evident that marked initial increase of P_a takes place, followed
by steady growth. Pool P_n, under similar conditions, suffers a
small initial reduction, followed subsequently by a rapid increase (curve "1.25"), which is more rapid than normal growth
(curve "1.0"). It is evident that the increase of the amino acid
pool is also very effective in increasing the nucleotide pool.
The reduction of the k_{20} by 25% has basically a reverse effect.

There is an initial decline of P_a. After reaching a minimum concentration, there is a very slow increase (curve ".75"). However, pool P_n, after trivial initial transient, starts to grow at a slow rate. These experiments show that pool P_n is strongly coupled with pool P_a. Figure 61 shows the effect of a 30% increase of k_{20} on enzymes and transport RNA C. It is evident that an increase of k_{20} also produces an increase in enzyme synthesis. However, transport RNA synthesis is initially reduced, but it increases later in a manner similar to enzyme synthesis. The reduction of transport RNA C, which occurs immediately after the increase of P_a, results mainly from the complex $[C\ P_a]$ formation. It appears that the increase of the pool P_a produces an increase in growth rate of all three entities. Computer observations revealed that all functional

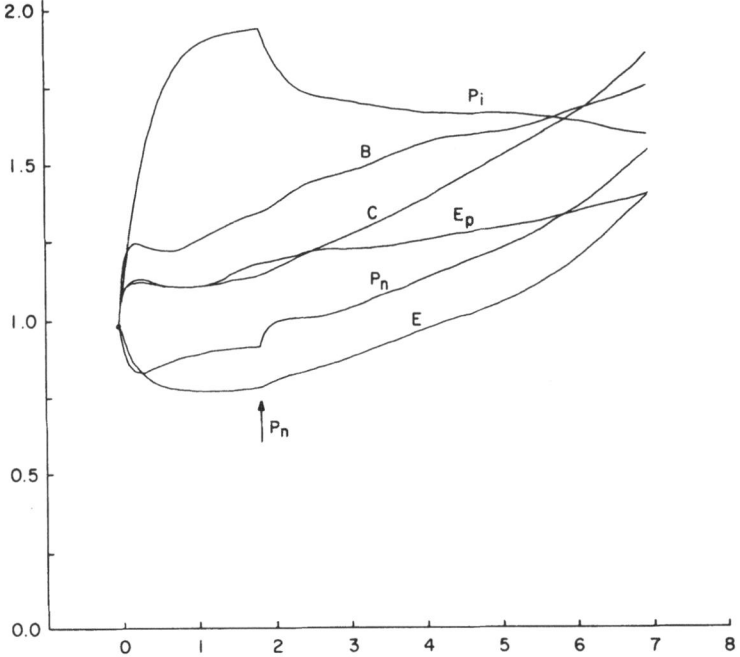

Fig. 59. The effect of increase of pool P_n concentration by increasing k_{19} from value 1.00 to value 1.25 at time indicated by arrow on: Pools P_i and P_n, enzymes E and E_p, ribosome B, and transport RNA C.

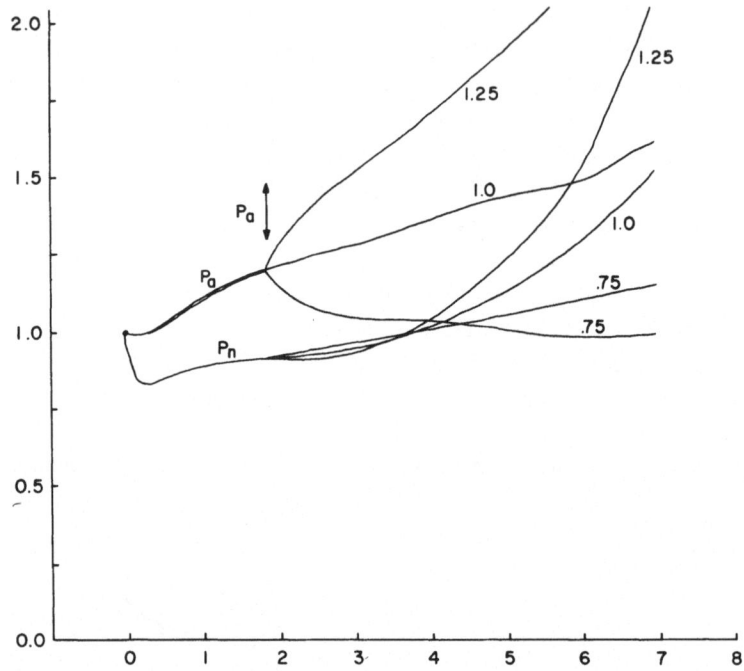

Fig. 60. The variation of pool P_i conversion into pool P_a. The effect on: Pools P_a and P_n. At the time indicated by arrow, rate constant k_{20} changed $\pm$ 25% from value of 1.00.

entities (after initial transient) also increased. It appears that an increased growth rate is associated with the increase of the protein pool, but an excessive increase of P_a leads to unbalanced growth. The general conclusion which can be drawn here is that conversion of pool P_i into P_a is a much more effective means of controlling the cell growth than conversion of pool P_i into pool P_n. This supports the previous conclusion that protein synthesis is a more effective means of controlling growth than nucleic acid synthesis.

c. Modification of Synthesis at Various Sites

The synthesis of messenger RNA represents a very effective means of controlling specific protein synthesis. This control is exercised on the genetic level. It can be increased, for example, by induction, or decreased by repression. Analysis of the repression–induction mechanism has been specifically discussed by us in different publications [1, 2]. No attempt is made here to consider this problem in detail. However, we should like to analyze the effect of the messenger synthesis on the overall cell growth. Here the messenger synthesis is considered to be an aggregate property of all messengers involved in protein E synthesis. Schematics of messenger formation are presented in Table II, equation 6. Here G_E interacts with the complex $[E_pP_n]$, and, as a result,

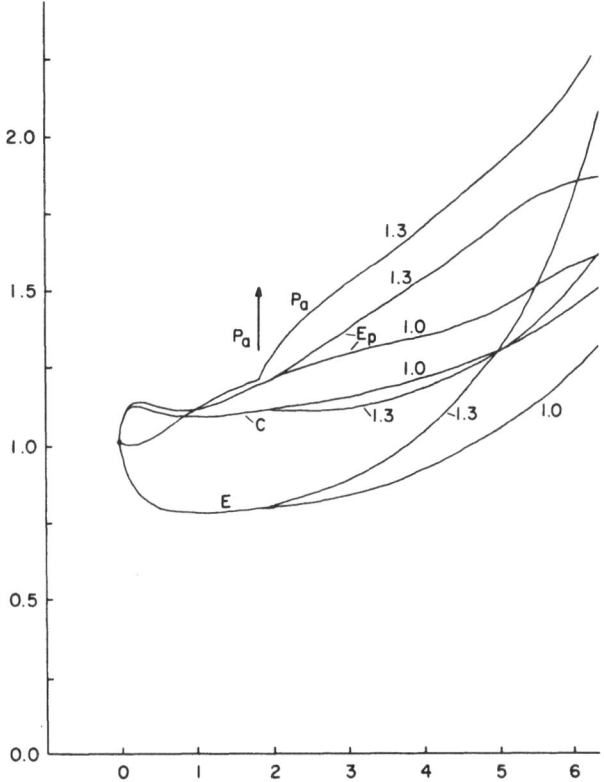

Fig. 61. The increase of pool P_i, conversion into pool P_a. The effect on: Pool P_a, enzymes E and E_p, and transport RNA C (rate constant k_{20} increased 30% at time indicated by arrow).

enzyme E_p is liberated and messenger M is produced. The equation for messenger M is presented in Table III, equation 7. It is evident that the term $k_6 G_E[E_p P_n]$ is the generating term for the messenger formation. Other terms are either for the utilization of the messenger or for the liberation of the mes- senger from a functional complex. It is clear that the rate of messenger M formation can be controlled by rate constant k_6. Modification of k_6 in the model-system would simulate the biological experiments where modification of genetic activity is produced by various agents, such as hormones, substrates, and antibiotics. For example, very selective inhibitors for RNA synthesis are various actinomycins [11]. It has been experimentally demonstrated that actinomycin D complexes with the DNA *in vitro*. This property of actinomycins has been useful as a selective tool in studies of messenger RNA syn- thesis. Simulation of these experiments which deal with RNA synthesis in cellular systems is of great importance.

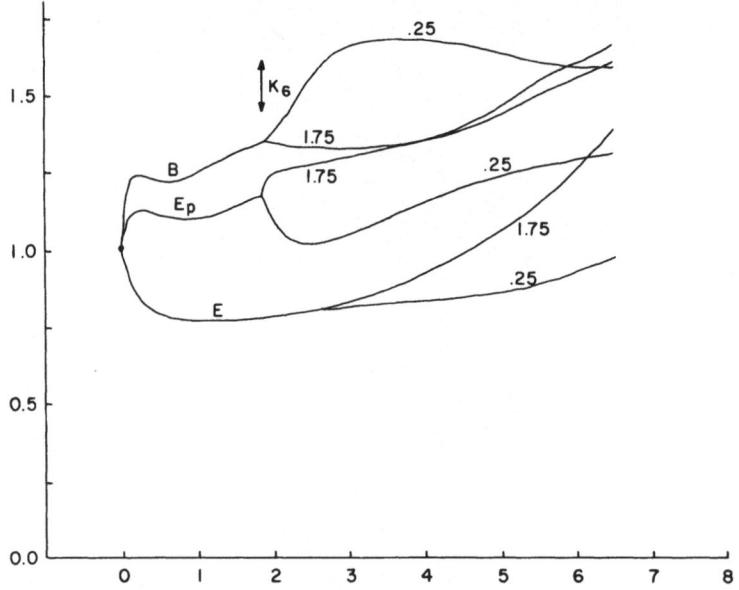

Fig. 62. The effect of the rate of messenger M formation on: Enzymes E and E_p and ribosome B recorded (rate constant k_6 varied $\pm$ 75%).

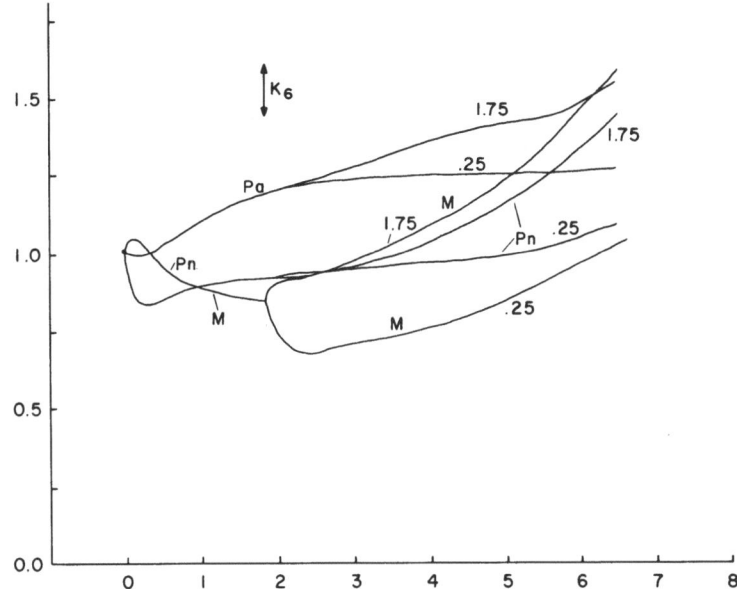

Fig. 63. The effect of the rate of messenger M formation on: Pools P_a, P_n, and messenger M recorded (rate constant k_6 varied $\pm$ 75%).

The question can be posed: what happens to the growth of the model-system when wide-range gene activation and gene inhibition occur? For this purpose, experiments were performed in which the rate constant k_6 was varied ($\pm$ 75%). Initially, the model-system was growing normally, and then, at a certain time, rate constant k_6 was suddenly changed. Figures 62 and 63 illustrate such an experiment, in which several principal functional entities are presented. Ribosome B concentration increases in a uniform manner until k_6 is suddenly varied. When k_6 is reduced by 75%, there is an immediate increase of ribosome synthesis (curve ".25"). However, the increase of ribosome concentration is only temporary. It reaches a plateau and finally starts to decrease. In contrast, k_6 reduction will immediately produce a reduction of free E_p polymerase (curve ".25") concentration, which passes through a minimum and then starts to increase at a slow rate. The principal reason for the reduction of free E_p is the fast buildup of complex $[E_p P_n]$, which is maintained at a high level, since k_6 is required for dissociation of the complex in messenger M synthesis (equation 6, Table II). For the enzyme E the effect is quite slow, but general growth rate is considerably reduced (curve ".25"). This is in contrast to the messenger M concentration which drops rapidly in the beginning (curve ".25"), but starts to grow later. In contrast, when k_6 is suddenly increased (75%), it is evident that most entities reveal a general increase of growth. However, ribosome B concentration initially suffers a reduction, but subsequently starts to increase again. The relationships between functional entities are too complex for detailed analysis. The following general conclusion can be drawn: the control of cell growth on the messenger level is not uniform.

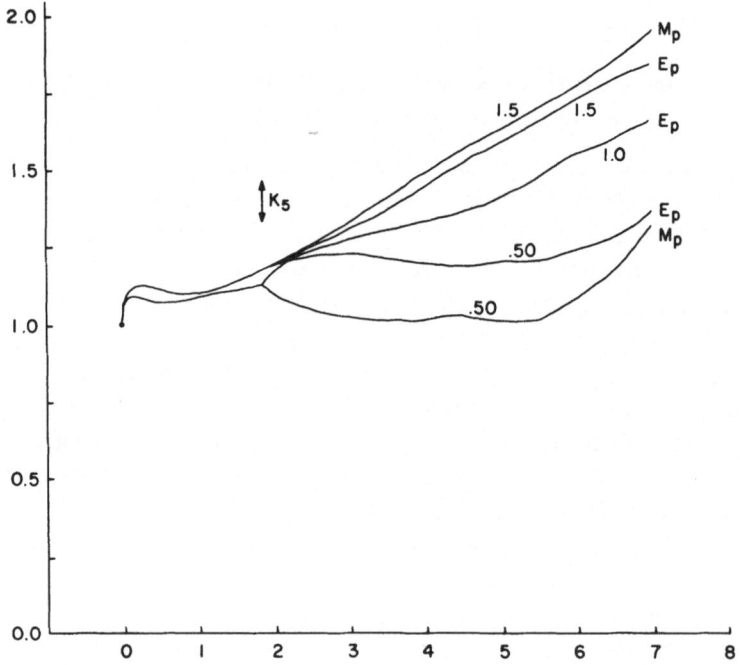

Fig. 64. The effect of messenger M_p formation on: Enzyme E_p and messenger M_p recorded (k_5 varied $\pm$ 50%).

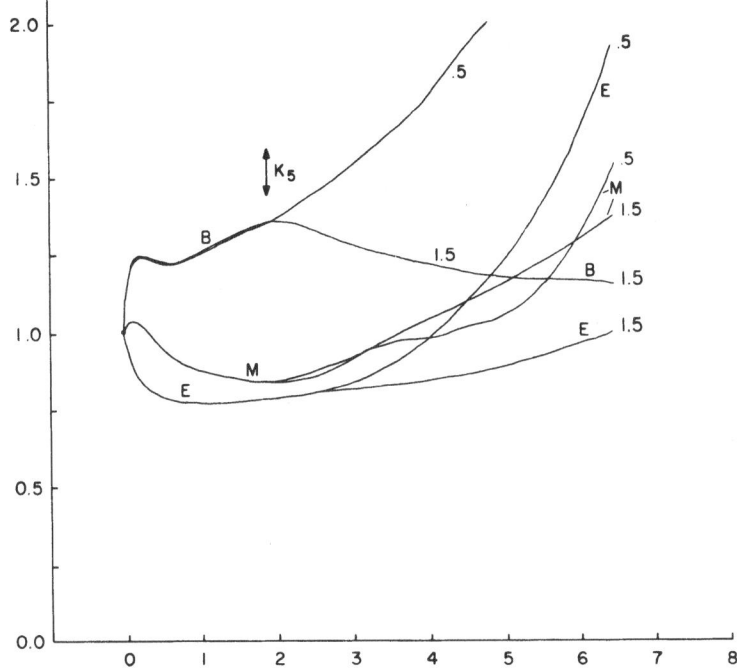

Fig. 65. The effect of messenger M_p formation on: Messenger M, ribosome B, and enzyme E recorded (k_5 varied $\pm$ 50%).

In order to study growth regulation at messenger M_p level, rate constant k_5 was varied. Figures 64 and 65 show some of the effects on some of the entities, when, at a time indicated by the arrow, k_5 was either increased or decreased (50%). It is evident that direct control can be exercised on E_p polymerase and messenger M_p concentration via k_5. Ribosome B, messenger M, and enzyme E concentrations indicate different trends, as shown in Fig. 65. Interrelationships in quantitative aspects are not at all clear. This can be especially observed when the kinetic behavior of messenger M is followed during the generation time. It is obvious that changes that occur at polymerase messenger M_p level produce unpredictable concentration variations in other functional entities. Furthermore, it appears that varying the polymerase formation does not constitute an effective means for controlling the growth rate of the model-system.

Ribosomes are basic functional entities required for protein synthesis. While there are a large number of messengers in the cell, there is probably one or only a few basic types of ribosomes capable of synthesizing all kinds of enzymes. Ribosome is basically formed from RNA and protein. The question will arise as to whether it is possible to control growth on the level of the ribosomal RNA or on the level of complete ribosome. Table II, equation 3, indicates that the RNA fraction of ribosome B' is produced when gene G_B interacts with pool P_n.

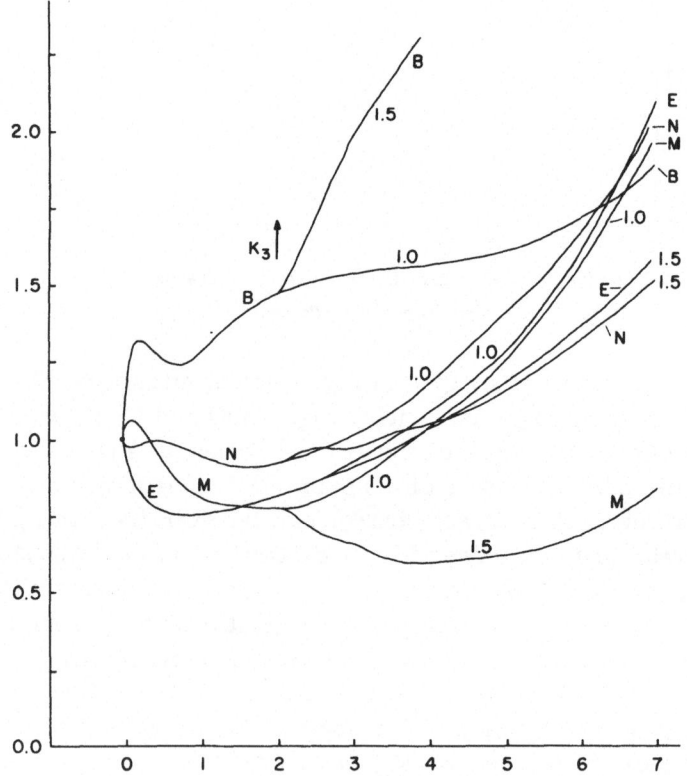

Fig. 66. The effect of B' formation on: Messenger M, template N, ribosome B, and enzyme E recorded (k_3 increased 50%).

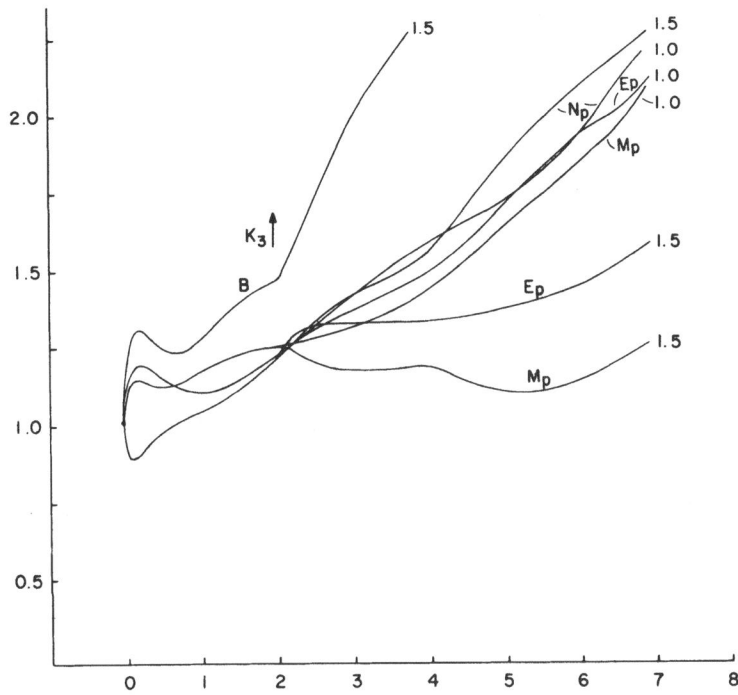

Fig. 67. The effect of β' formation on: Ribosome B, messenger M_p, template N_p, and enzyme E_p (k_3 increased 50%).

B' interacts with the protein fraction, as indicated in equation 2, forming the ribosome. Alterations in B' concentration can be produced by varying k_3. Experiments in which k_3 is increased by 50% at times indicated by the arrow reveal that drastic concentration changes occur in various functional elements (Figs. 66 and 67). As expected, the increase of the k_3 produces an immediate increase in ribosome B concentration. However, this growth is associated with the reduction of messenger M and M_p concentration. Figure 66 further indicates that the increase of ribosomes does not increase the synthesis, since there is a definite reduction of template N and enzyme E. Figure 67 indicates that while messenger M_p and polymerase E_p both are reduced, there is a slight increase in N_p concentration. Figures 68 and 69 show the results when k_3 is reduced

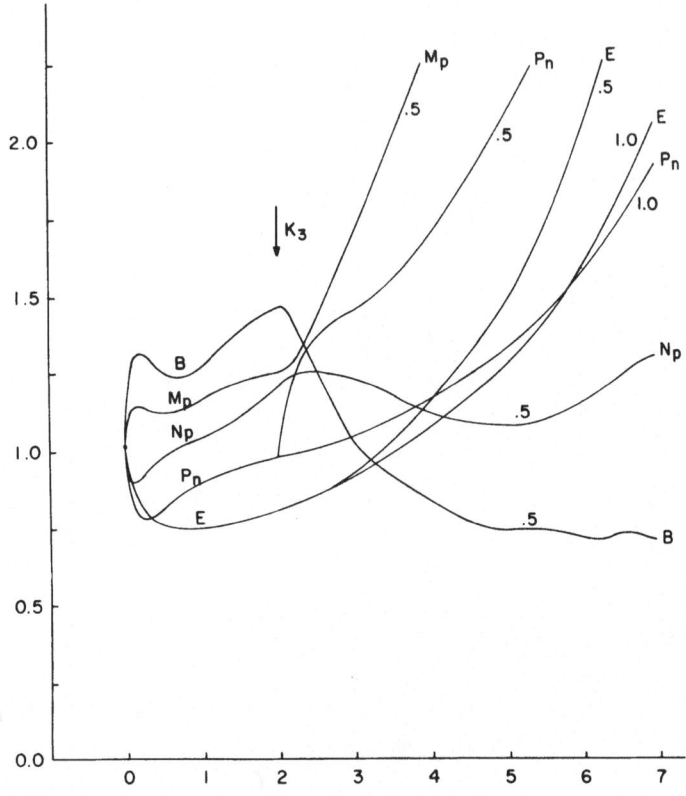

Fig. 68. The effect of B′ formation on: Enzyme E, pool P_n, template N_p, messenger M_p, and ribosome B (k_3 reduced 50%).

50%. The immediate effect of the reduction of k_3 is the drastic initial decrease of ribosome B concentration, but gradually only a slow decline occurs.   On the other hand, there is an increase in M, M_p, P_n, and E concentration when k_3 is reduced. It appears that by variation of the ribosomal RNA fraction, it is not possible to obtain adequate growth regulation. However, it is considered of interest to find out whether growth control could be obtained at the level of the ribosome itself. Experiments were carried out on the computer (equation 2, Table II). Results indicated a very poor growth regulation when constant k_2 was varied.   It appears, in general, that neither the RNA fraction nor total ribosome syntheses has effective regulatory characteristics which are desirable for uniform control and regulation.

Another basic functional entity involved in cellular growth is transport RNA. Since it participates directly in protein synthesis, it is worthwhile to analyze briefly its role as a regulatory agent. As indicated in Table II, equation 4, the rate constant k_4 controls RNA C synthesis, and by varying k_4 it is possible to vary the rate of C synthesis. Experiments were carried out, but the results obtained were of limited significance. Figure 70 shows that after the reduction of k_4 for 70% some very drastic changes occur in the concentrations of

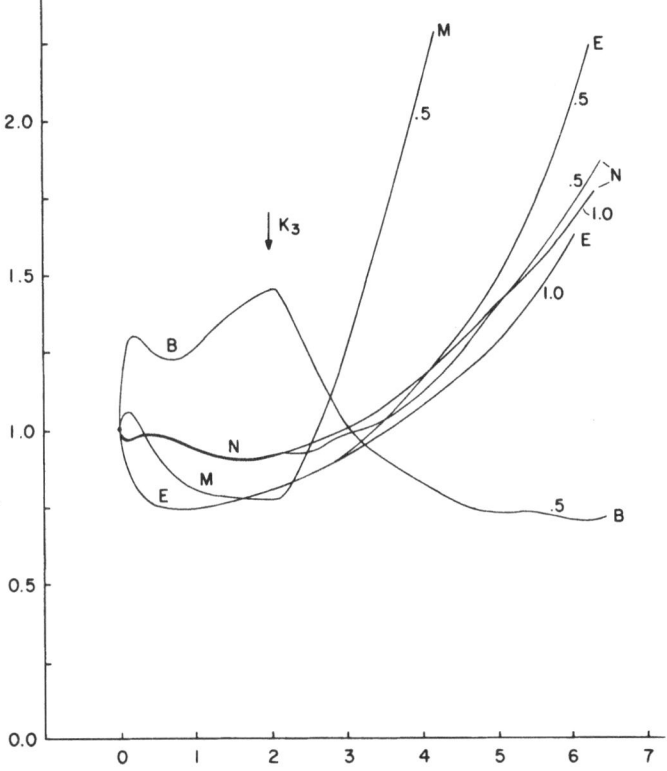

Fig. 69. The effect of B' formation on: Enzyme E, template N, messenger M, and ribosome B (k_3 reduced 50%).

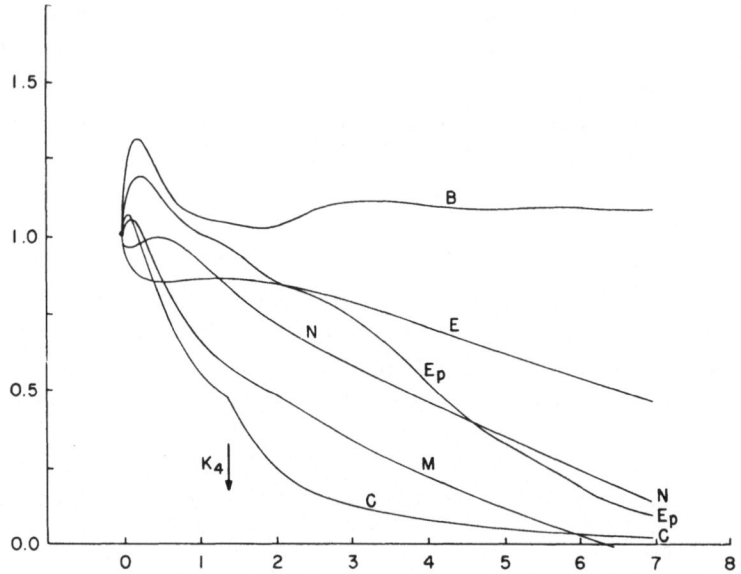

Fig. 70. The effect of transport RNA C formation on: Ribosome B, enzymes E and E_p, messenger M, template N, and transport RNA C recorded (K_4 reduced 70%).

various functional entities. First, there is a rapid reduction of C. This is associated with the reduction of messenger M, template N, and enzymes E and E_p. Ribosome B concentration increases slightly, and it is maintained at this level during the generation time. This drastic change in k_4 of course seems to lead to the destruction of the model-system, because the concentrations of all functional components are declining. However, when changes in k_4 are less drastic, there is correspondingly a slower decline in the growth rate. It appears that growth regulation could be exercised at transport RNA level, but it is not as smooth an operation as protein synthesis controlled directly on the template level or at the pool level. The very fact that ribosome concentration does not behave as other entities do (Fig. 70) indicates that on the gene level there is competition for pool P_n. It was further evident that the control mechanism was not very sensitive or uniform. For example, the increase in the value of k_4 had a very sluggish effect on the growth of the system. It appears that quantitative features of the model-system are such that transport RNA C is not a limiting factor. Since quantitative interrelationships between various functional entities and rate constants determine the characteristics of the model-system, it seems that by certain modifications of rate constant values it would be possible to arrive at a system which would exhibit better control at transport RNA level. But the very fact that such a regulatory operation is highly dependant on detailed quantitative properties suggests that it is not very effective or safe as a general regulatory mechanism. More specific analysis at this point should perhaps be carried out in the future. This is especially so, since it has been suggested in the literature that transport RNA may play a leading role in cell growth regulation [3]. In summary, it appears that cell growth can be increased and decreased by regulatory processes operating at various levels in functional organization. However, the most desirable type of mechanism of regulation is one in which action will produce a uniform and synchronous change in all functional entities. Model-system studies reveal that factors which directly influence protein synthesis will effectively control overall cellular growth.

7. CELLULAR INJURY AND DEATH

a. Introduction

The phenomenon of cellular injury and death is a very perplexing problem in nature and has many ramifications. It is a problem of great intellectual challenge as well as practical importance. It has been a subject of much speculation and many theories have been proposed to interpret the mechanism of cellular death. Before we analyze this problem in more detail, it is important to consider the viability problem in general and to explore its practical aspects. Historically, the concept of cellular death has been a subject of mystical speculation as well as arbitrary, oversimplified mathematical formulations. It is essential to put the concept of cellular viability on a rational basis, and one should attempt to interpret cellular injury and death via basic metabolic and synthetic processes.

As a matter of fact, cellular death is a far simpler phenomenon than cellular growth and development. The growth process is a complicated sequential interaction between various functional entities, and there is a high degree of order in such a process. By contrast, cellular death can be initiated by a relatively simple event followed by highly complicated disorder. Furthermore, one cannot consider cellular death only from the point of view of complete and absolute killing. There are degrees of injury, so that there may be as a consequence cellular recovery. In principle, there is no absolute concept of cellular life or death. There do exist states of functional disorganization, and these states can be altered by internal or external factors.

There are many practical implications in the phenomenon of cell injury and death. Let us review first what we call selective killing. This is based on the concept that various cells have different biochemical and physiological characteristics, which permits us to apply some agents to inactivate selectively one particular cell population while leaving other cells intact. This type of selective killing of cells is important for the process of therapy, where it is imperative that the basic multicellular system maintain its integrity while one is able selectively to destroy other parasitic cells operating in the same system. For example, we may try to kill off the bacteria which have invaded a multicellular organism, or we may attempt to destroy some abnormal cells which have arisen within the system. In order to carry out selective killing really efficiently without injuring the parent system, one has to have great insight and information as to the differences between parasitic organisms and the parent organism, in which the parasites operate. At the present time this type of selective killing is possible between cells which have distant genetic origin. For example, bacteria cells can be, in most cases, relatively easily killed in the presence of mammalian cells. On the other hand, a cancer cell is very difficult to kill selectively in the presence of normal cells of the same origin. One would expect that only when we obtain a relatively wide spectrum of information in regard to metabolic systems of different cells will it be possible to apply selectively agents which inactivate or injure only one type of cell in the presence of many other types of cells. For this reason, it is not sufficient to regard the viability problem from a general phenomenological point of view, although it is essential that detailed specific information of all systems be available. Successful therapy is based on the knowledge of differential biochemistry, physiology, and morphology of individual cell types. At the present time, we do not have this knowledge, and consequently therapy at the present time is highly empirical.

Microbiological sterility is another area in which cellular injury and growth play important roles. The sterility problem can be very simple in some cases. For example, in the sterilization of equipment and certain supplies, where there is no differential effect, the sterilization procedure can be straightforward. However, quite often it is necessary to sterilize under conditions where the basic medium in which the contaminating cells exist should not be altered at all or altered only by a minor degree. This requirement sometimes makes the process quite difficult. Sterilization of foods or pharmacological supplies by various methods such as thermal, chemical, and radiation processing requires that the basic item be very little altered by the sterilization procedure, while cells or spores should be killed. This is by no means an easy requirement, because large doses of radiation or heat required for sterilization often also alter characteristics of the basic medium. Here again, basic knowledge in spore physiology and cell physiology is essential for developing proper techniques for the sterlizing process. Other types of differential requirements occur, for example, in the sterilization of vaccines, where it is essential that antigenic characteristics be left intact, but infective properties be destroyed. Here again, a selective interaction with sterilizing agents which will affect one system but not the other is essential. Many vaccines have been produced in which cells have been effectively killed, but the antigenic properties have been so tremendously reduced that the vaccine has had only minimal therapeutic value.

Growth interference and cellular injury can be caused by agents which exist in our normal environment: for example, air pollution, high frequency electromagnetic fields, ionizing radiation, etc. While the level of cellular injury occurring in these conditions may be low, however, since exposure may be continuous, this type of injury may induce some basic changes in cellular characteristics. Here again, it is essential that the biological system be well understood before any effective protection can be found besides removing the agent.

In general, a cell can be killed or injured in a multitude of ways. It can be killed by radiation, by excessive temperatures, by mechanical injury, by chemical agents, etc. This raises the question of whether the process of death is basically a singular process, or is a diverse and complex set of disintegrative phenomena. It would be of benefit to review briefly the basic views existing at the present time on this subject in the literature. Many theories have been proposed to interpret the cellular killing process, but of the foremost importance from the view-point of popularity is the so-called target theory. It is based essentially on population statistics where the cellular inactivation or killing rate will provide information in regard to the killing process, which may be a single-hit or multihit type [12, 13, 14]. Despite its obvious shortcomings, the target theory is still widely accepted as a means of interpreting and elucidating the mechanism of cellular death [15, 16, 17, 23]. Since target theory is based on the curve form analysis, which itself is affected by a multitude of experimental parameters [24, 25], one cannot expect that the "death kinetics" of a population will yield information with regard to the cellular killing process and mechanism of cellular recovery on a single cell basis. In the literature numerous curves have been published for such analysis; no significant results have been obtained. One could ask whether the theory is founded on the premise which potentially could give information for cellular death mechanism, or are we dealing here with an obscure illusion?

The primary source of information for target theory is the relationship between the rate of killing and the exposure dose.

In our opinion, this type of relationship does not contain any basic information which is necessary for interpreting cellular killing mechanisms. This can perhaps be best represented by the following example. Let us assume that an experimentalist is the gateman for a cemetery in which the burial of the "dead" takes place. One should be able to ask whether this gateman is able to diagnose the cause of death from the rate at which the coffins arrive at the cemetery. Obviously, there is no relationship between the cause of the deaths and the rate of the arrival of coffins. These are independent entities, and cannot be submitted for analysis of a mechanism which constitutes an interaction between those two entities. The gateman never knows from the primary point of view what is the cause of death when the coffin arrives. The arrival rate may depend on many external factors for which there is no information for a gateman. The question can be asked: Why did target theory find such wide acceptance among biologists, who were well aware of basic biological processes, in spite of the fact that during the last few decades, during which time target theory was highly popular, no significant results had been produced? To the contrary, it appears that a large amount of fruitless effort and labor for many scientists has been lost. How could this type of situation arise? It seems that one of the reasons for this state of affairs is the fact that the majority of biologists are not competent in mathematics. Target theory, which basically represents a rather simple application of statistical analysis, had some effective window dressing to scare the biologists from really looking into the basic foundations of the theory. Consequently, this type of theory was accepted on a mass psychosis basis. In addition, the target theory found wide acceptance by physicists who had migrated into the field of biology. Many of them did not have the proper biological background and helped to elevate a highly arbitrary theory to popularity.

A few decades ago it became obvious to some that target theory was capable of explaining neither cellular injury nor cellular death. The problem was discussed during various scien-

tific meetings and heated arguments developed on many occasions, but no substitute for target theory appeared for a while.
Then some new proposals were made by us for the interpretation
of cellular injury and death phenomena [18]. Later, a wider and
more comprehensive treatment of the subject was made [19].
The basic premise proposed was based on the concept that there
is a definite relationship between cellular injury and functional
disorganization of the metabolic system.   One could say that
the factors that interfere with or excessively modify cellular
organization and cellular processes can be considered in a
biological sense to injure or kill the cell. Such interference
could result from the interaction of cellular functional entities
with a multitude of agents, for example, absorption photons,
collision with elemental particles, interaction, with various
molecules, etc.   As a result of these interactions, in terms
of functional subunits of the cell, the following changes could
take place: inactivation or alteration of enzymes; modification
of messenger and transport RNA; inactivation of genes; alteration of metabolites, intermediates, and cofactors.   Since
various agents and molecules can interact differently with
cellular functional entities and molecules, a multitude of modes
of functional disturbances may develop, depending on the character of the agent involved. It can also be considered from a
chemical point of view that the molecular alterations produced
within the cell may be reversible or irreversible. Furthermore, metabolic damage produced in the cell can be repaired
by the cell, or, if this is not possible, the cell will die. However, at certain levels of injury the cell may survive when it
can compensate for the deficiencies itself, or when external
agents are applied which help cellular recovery. Consequently,
we can define cellular injury in general as a phenomenon in
which normal functional processes of the cell are altered by
interfering agents.

Since various operational entities of the cell have specific
molecular composition and organization, the chemical interaction specificity determines which of the operational units
are altered by a particular agent. The specificity of the agents

may determine their character of functional injury. This can be a singular injury in single locus, or it can be a multiple injury in many loci. For example, injury on a single gene level would be sufficient to kill the cell if this gene is irreversibly inactivated. By contrast, when several of the enzymes are inactivated, we have widespread functional damage, and it is not at all obvious how the cell would be affected by such injury. As a matter of fact, cellular processes are so complex and so interwoven that it is very difficult to perform a clean-cut analysis of the system, unless the system is analyzed quantitatively. Therefore, it is essential that formalized relationships be established between various functional elements of the system, and that the problem be solved in a quantitative manner. It is of interest that the cellular model-system which has been developed can serve an additional useful purpose to that of studies on growth regulation and growth in general: as a means of analyzing the problem of cellular viability.

In order to gain more insight into cellular disorganization, it is essential that a systematic study be made on the effects which can be produced by interfering with the activity of important functional entities in the cells. Furthermore, it would be important to view the problem of cellular disorganization not only from a phenomenological point of view, but also to study the level and degree of disorganization. As it appears later, the stability of functional elements is an important factor in determining the character and the depth of the cellular injury. All these problems will be analyzed subsequently in the quantitative manner using the established operational model-system. Occasionally some parameter modifications will be made in the model-system in order to facilitate computer analysis in some particular type of experiment.

b. Basic Premises and Procedures

Quantitative studies on the cell growth model reveal that the system is sensitive to the alterations which occur in any functional entity, and that the change which occurs is followed by adjustment at all levels of functional entities in the model-system. Therefore, in order to understand the events that are taking place after the interaction at a particular site, it is essential that all changes occurring in other entities be recorded simultaneously. Such computer study is equivalent to the study of cell physiology under conditions in which, simultaneously with the cellular injury, physiological and biochemical measurements are made with all possible entities which enter into the scheme. This is essential, since the measurements of a single entity are not sufficient to characterize the kinetic events in the cellular metabolism after a certain type of injury. However, in many practical conditions in cellular biology, injury occurs simultaneously at multiple sites. This is especially true in cases when a heterogeneous agent is introduced into the system, such as radiation, heat, etc. No attempt is made to explore the multiple injury pattern, not because it would be too complex to be analyzed, but rather because it is too involved and would require a special programming for that purpose.

As will be seen, the model-system enables us to study specific injury where various drastic changes may occur in the concentration of functional entities. Those changes may not be uniform and may depend on the type and locus of the injury. It can be such that drastic changes occur in some components, while other components are affected only slightly. In these conditions, one will get an abnormal concentration pattern of cellular functional elements. Such computer experiments can be compared in radiation biology to the so-called "monster" cell formation. These cells do not have normal composition, functional behavior, or appearance.

What are the basic premises which will permit us to analyze the cell model-system for the studies of cellular injury and to derive conclusions which could be projected to interpret injury mechanisms in actual cells in the living state? All basic entities in the model-system presented in Fig. 1 are part of a living system. They represent major parts (volume and weight) of the functional entities of the cell. It appears that it is possible to analyze the effects of various agents of cell viability by studying group genes or total protein or total transport RNA, etc. Regulatory mechanisms and steps in the living cell are not well known, but these are complex, and direct analysis of composite mechanisms is not possible at the present time on a model-system basis. However, the result of the interaction of an agent with a regulatory unit, which is part of a mechanism, will appear as a change in the concentration of a major functional element(s). This is so because regulatory elements control the operation of the basic functional entities. One may add that simulation on the model-system need not follow the exact kinetics of the process occurring in the actual cell where the number of specific functional elements is large, but nevertheless basic phenomena of disorganization can be demonstrated. One of the principal reasons why model-system response is kinetically more drastic is the competition effect which occurs in the utilization of various pools. The response in the model-system may be kinetically exaggerated.

In the analysis of cell growth kinetics we use as the time base the "generation time," during which the concentration of functional entities is roughly doubled. In order to analyze the kinetics of cellular disorganization, it is essential that the observation time be extended. In our experiments, we have varied the extended computer time scale between five to nine times the normal "generation time." Selection of the proper time scale depends on the particular type of experiment. In general, the observation time has to be long enough to resolve the problem of cellular recovery and cellular death which may take several "generation times." For many studies, it appears that after the initial decline the model-system may be able to recover, and in order to solve this problem, observation time has to be long enough that concentrations of all functional entities can be followed. Sometimes there is a loss of most of the principal entities, but a few entities are left in the system at low concentrations at the end of observation time. Repeated computer studies reveal that no single entity alone is able to restore functional characteristics of the model-system and thus produce growth. Simulation of cellular experiments on the computer requires that we be able to introduce and remove the agents which interact with a particular functional entity. This is accomplished on the computer with the aid of synchronized electronic switches. At any time, an agent at a specific concentration can be introduced into the system and again removed. Consequently, we are in a position to control the concentration of external agent at the active site, and we can control that agent all of the time during the experiment. This is a highly flexible experimental arrangement and can yield a variety of information. Unfortunately in practical experiments in biology, it is not possible to observe all of the functional entities in a single cell while the experiment is carried out, and this drastically limits the amount of information to be gained from a single experiment.

In previous studies on cell growth regulation, the experiments were performed under conditions in which the growth rate (in a positive or negative sense) changes were not too drastic. However, viability studies will be carried out under conditions in which the action of the various agents can be extremely drastic, even to the level at which the model-system becomes nonfunctional. Nonfunctionality of the model-system is equivalent, in terms of biology, to the loss of cellular viability. Complete disorganization of the model-system and loss of functional entities represents a "nonviable" system.

Introduction of very drastic parametric changes into an active model-system requires that the system be capable of self-organization and self-recovery. Simulation of cellular injury and "death" processes on the model-system requires that that system be "disturbed" at the level of various functional entities. Sometimes disturbing stimuli have to be very strong. For this reason the model-system was partially reorganized in the sense that regulatory features were made stronger. It was considered that these characteristics would help the cells sustain the injury better and also help cellular reorganization after the injury. For this purpose, the values of the rate constants k_{21}, k_{22}, and k_{23} were increased, while reversible rate constants k_{-21}, k_{-22}, and k_{-23} were left the same as before. The effect of regulatory compensation will be analyzed later in various experiments.

How the cellular injury is simulated on the model-system will be presented at the time when specific effects are studied. However, in general, an injury can be represented basically as a permanent or transient increase of a certain entity or rate constant. The nature of the experiments depends largely on what kind of cellular injury we desire to simulate. For example, we can expose cells to an agent while growing continuously, then terminate the exposure, and observe the effects in the growing state. Such an experiment would simulate the action of a drug which we add to the growth medium, then remove, allowing the growth to continue. The simulation of the injury process on the computer is rapid. A change in rate constant, in kinetic terms, is represented by a rectangular impulse. This would be, for example, equivalent to rapid neutralization or removal of the agent. However, since some chemical agents penetrate the cell much more slowly, one should consider that a gradient-type simulation would be necessary. However, at this level of affairs, such refinement in kinetics is not essential, since we are mostly interested only in the basic effects.

The view during the early days of target theory was that when a cell was killed it was dead, and irreversibly so. Experimental evidence contradicted this type of pragmatic view. It was especially evident from radiation experiments that various cells were capable of recovering depending on their post-injury condition and treatment [20, 21, 22]. No attempt will be made to review the development of this field here. However, a few basic considerations which have been presented previously [19] are worthwhile for analysis. It is expected that a cell will recover from an injury when the exposure time is not too long and the degree of injury is not too extensive. Simulation of experiments via the model-system require that we specify what we mean by recovery. Since in this book we treat cell growth problems only in a quantitative manner, we do not analyze genetic division from the point of view of injury. However, we analyze the process of synthesis of all essential functional entities which are required for cell growth. If a cell has suffered damage but is still capable of re-establishing its synthetic functions, it will be considered that that cell is able to

recover.   There are various modes and levels of injury. The
simplest type of injury one may consider is a reversible type
chemical reaction. Recovery from injury means that the chemical reaction at some particular site is reversed by another
chemical which is introduced for this purpose. A more complex reversible type of growth inhibition can be produced in a
cell by one hormone and reversed by another. However, in
this case we have a reversal process, and not a single reaction step.   From the point of view of cellular damage, it is
obvious that cellular recovery would be perhaps easiest when
there is only a single-step type of damage. However, even
here there are certain time limits as to how long the cell can
sustain that type of injury. When a cell is injured extensively,
especially by an irreversible chemical type of injury, none of
the aforementioned means are adequate to restore cellular
viability.   On such an occasion it is essential that specific
synthetic processes be induced to restore normal physiological
processes.   What are the requirements for cellular recovery?
First, a cell should be able to perform certain basic synthetic
processes.   Second, the cell should possess a small pool of
metabolites, which could be utilized for minor synthesis.
Third, environment and temperature should be suitable for
synthesis and metabolism. Under extreme conditions, when
cellular injury has imbalanced intracellular organization in
terms of the concentration of functional entities, it may be
essential to inhibit some phases of synthesis. However, under
conditions where the cell is not capable of carrying out all
necessary synthesis, it may be necessary to supply externally
certain metabolites and intermediates.

Since all functional entities represent group properties in
the model-system, simulation of cellular injury and recovery
will be represented by a single-step type of reaction. The
recovery process is followed after the removal of an agent as
a function of time. When "injury" of the model-system is too
severe, agents are introduced into the system which help to
restore it to the functional state.

c. Genetic Injury

Basically, genetic injury can be reversible or irreversible. In irreversible injury, gene activity is partially or completely destroyed, and this effect is permanent. Reversible injury is a type which can be corrected either by removal of the agent which interacts with the gene or by the action of repair mechanisms. For example, if there is a complexing process between an agent and the gene, then the [gene-agent] complex can be maintained as long as the agent is present in the system. However, when the agent is removed, the complex dissociates and the gene is free to operate again. Another aspect of genetic injury is that it can be either partial or complete. In the latter case, the gene is completely blocked and is not functionally operative. Partial injury means that the gene is operational, but it does not function at the normal rate. This condition can arise when the spatial configuration of the genetic structure suffers minor modifications. In partial genetic injury, the average genetic activity is reduced. Furthermore, we assume that when the gene is blocked it is not operational, and does not produce RNA. We do not analyze the case where the gene is producing a defective, nonfunctional RNA. This can be done, but it requires some special programming and model-system changes.

Genetic Injury at G_E Level . Gene G_E participates in RNA synthesis, and injury at that level will be analyzed from various points of view. The basic effect is considered to be the termination of messenger M formation when the gene is in the inactive state. Equation 6 of Table II represents the basic gene interaction in messenger M formation. This reaction yields the first generating term in the differential equation (Eq. 7, in Table III). It is evident that gene G_E activity depends on the rate constant k_6. We have decided in our experimental technique to modify the gene activity by varying rate constant k_6. It is also evident from equation 7, Table III, that other terms are not involved in messenger M formation. The normal growth of all the functional entities during one generation time is represented in Figs. 71 a and b. The quantitative values of parameters used in the reorganization of the model-system will be maintained throughout the viability studies, but occasionally some changes will be made, and these will be pointed out specifically in a particular experiment. The time when the transient rate constant changes are introduced into the system varies, but on most occasions it is roughly three-quarters of a generation time. Occasionally in special experiments there are deviations from that value. In simulation of cellular injury in Figs. 72–74, the time scale is extended to four "generation times." Figures 72–74 represent the experiment in which k_6 was reduced at the time indicated by the arrow. Genetic damage was only partial (genetic activity was reduced 93%). The reduction of activity was permanent, and growth was studied under these conditions. As a result of this genetic

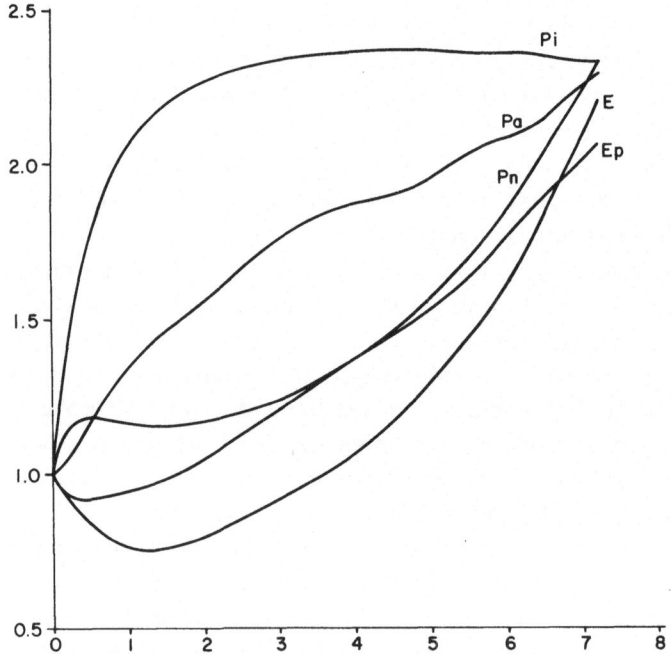

Fig. 71a. Concentration of enzymes E and E_p, pools P_i, P_a, and P_n during normal growth.

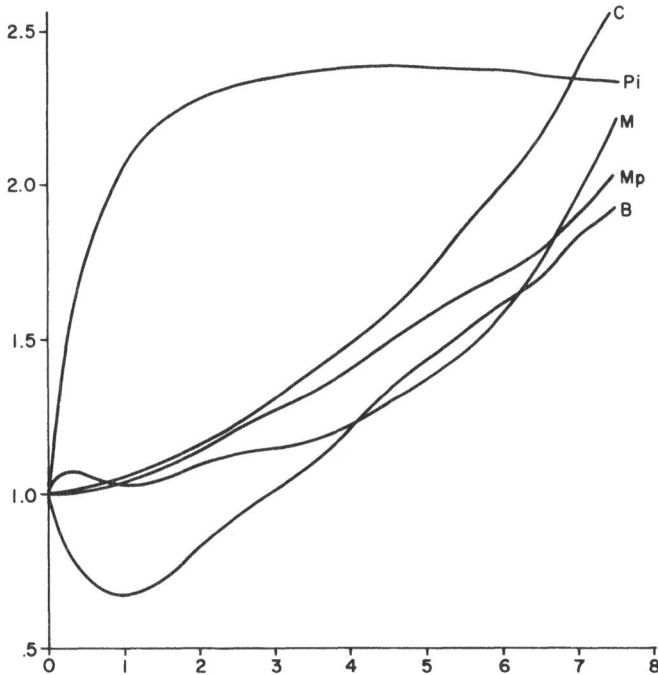

Fig. 71b. Concentration of messengers M and M_p, pool P_i, ribosome B, and transport RNA C during normal growth.

damage, some components of the system started to grow very rapidly, while others were drastically reduced. Figure 72 reveals that after introduction of the genetic injury at k_6 level there is immediately a drastic reduction of messenger M and template N concentration. While M continues to decline, N passes through a minimum and starts to increase again. In contrast, enzyme E continues to increase, reaches a maximum, and subsequently starts to decline. Introduction of a reduced k_6 value has no effect on transport RNA C growth, which proceeds rapidly. However, it reaches a maximum and subsequently declines continuously. Figure 73 shows that there is a rapid growth of ribosome B, which passes through a maximum and then continues to decline. Pools P_a and P_n continue to grow and both pass through a maximum. While pool P_n

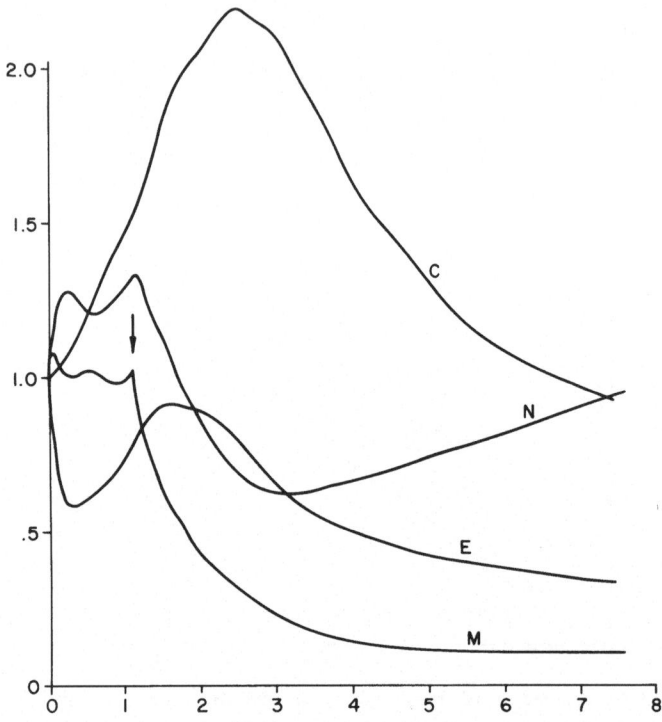

Fig. 72. The effect of genetic block at gene G_E level on: Messenger M, template
N, enzyme E, and transport RNA C.

starts to decrease, P_a passes through a couple of oscillations
and then starts slowly to increase. Figure 74 shows that E_p
polymerase continues to increase for a while, then passes
through a maximum, and finally starts to decline slowly.

It is of interest to note that the messenger M_p level is low.
This is mainly caused by the phenomenon of a large amount
of ribosome available and consequently a rapid formation of
template N_p. Template N_p is initially reduced, but after
passing through a maximum, a slow decline occurs. Since
Fig. 74 is recorded in reduced concentration scale, it can be
seen that drastic concentration differences exist between
enzyme E and enzyme E_p polymerase. If the growth of this
system is continued, observations on a more extended time
scale reveal that all the components of the system finally

decline, and the system ceases to exist.   In this system, in order to avoid computer overloading, rate constants k_{16} and k_{17} were increased, which means proteins were made more unstable, and, consequently, the growth rate of the cell was reduced.   This is evident when Fig. 71b is compared with initial growth in Figs. 72 and 74.   In summary, it is evident that a 93% inactivation at gene G_E level will produce drastic changes in the model-system and that those changes are highly irregular. In terms of normal cellular biology this type of injury would be characterized by the production of great morphological changes in the cell.   If the division mechanism of such cell is not impaired, its growth should continue for several generations, but the cell will finally disintegrate.

A recovery from cellular injury on gene G_E level is indicated in Figs. 75–77.   Here, at the time indicated by the first arrow, the k_6 again acquires its normal value.   The experi-

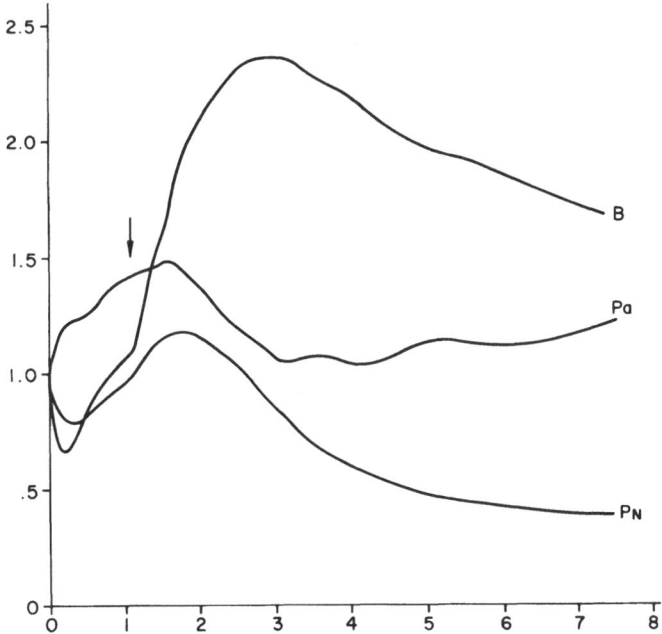

Fig. 73. Same experiment as recorded in Fig. 72. The effects of genetic block on: Pools P_n and P_a and ribosome B.

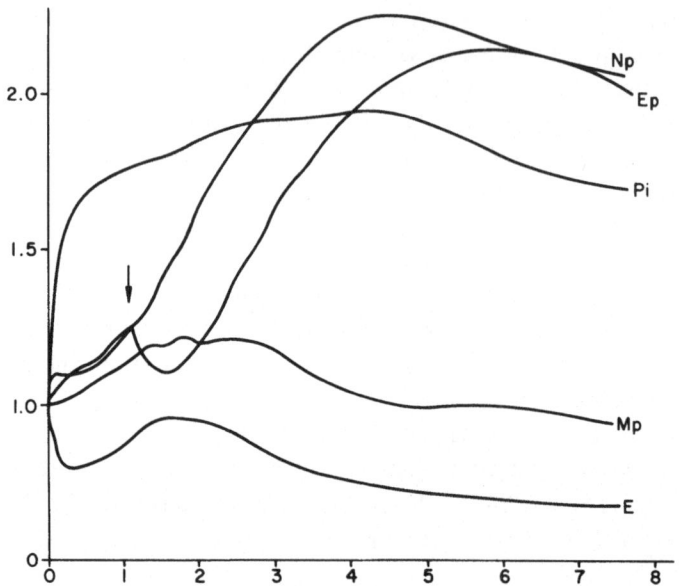

Fig. 74. Same experiment as in Fig. 72. The effects of genetic block on: Template N_p, enzyme E_p, pool P_i, messenger M_p, and enzyme E.

ment indicates that under these conditions the cell can recover rapidly, provided that the time is not too long for genetic repression.   When the genetic block is removed, there is immediately a rapid increase of messenger M formation. This M increase is rather dramatic.   It reaches a very high peak, is subsequently reduced to a minimum value, and then starts to grow again. After a small initial transient, all components of the synthetic system (M, N, C) of enzyme E increase rapidly.   At the same time, there is a dramatic reduction of ribosome B, but after passing through a minimum, B starts to increase also.   The principal cause of rapid ribosome reduction is the combination of rapidly formed messenger and ribosome into the template.   Figure 75 shows that there is indeed initially a very rapid template N formation.   This is followed by a slow phase, but finally a rapid increase of N again takes place.   When N growth reaches a plateau, M, E, and C start to grow rapidly.   Figure 76 shows that there is a rather rapid initial increase of M_p formation, followed by a period of constant concentration; subsequently, however, growth will be resumed. In contrast, there is a reduction of template N_p which, after passing through a minimum, starts to increase at first slowly and then rapidly. Figure 77 shows that all pools increase initially, but that they have different growth patterns. P_i shows a constant phase followed by a slow growth. Pool P_a starts to grow steadily, while pool P_n has a delay before rapid increase occurs. It is of interest here to note that during all this injury operation, internal pool P_i concentration changed comparatively little, in spite of the fact that the con-

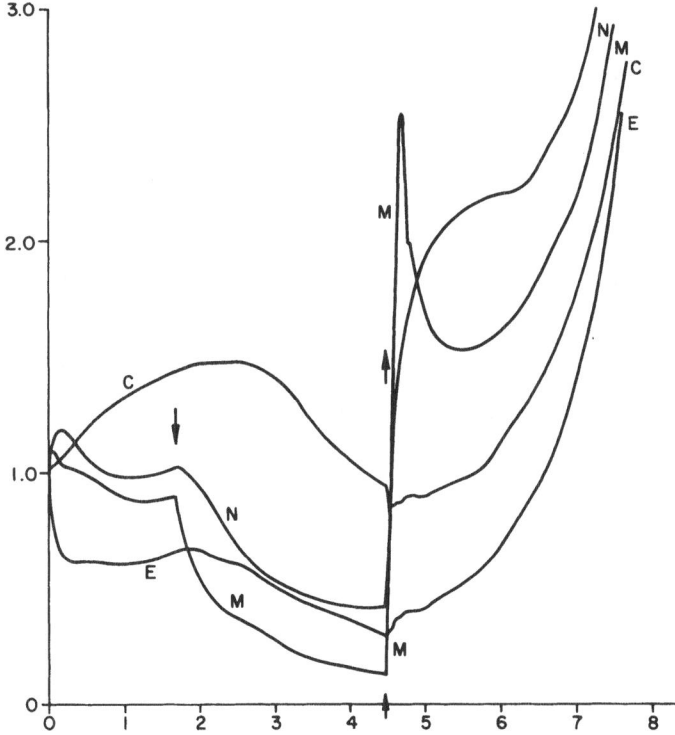

Fig. 75. Recovery after removal of genetic block. At time indicated by first arrow (↓), k_6 activity is reduced 93%. When genetic block is removed (↑), gene is again in normal functional state. Concentrations of template N, messenger M, transport RNA C, and enzyme E are recorded during the experiment.

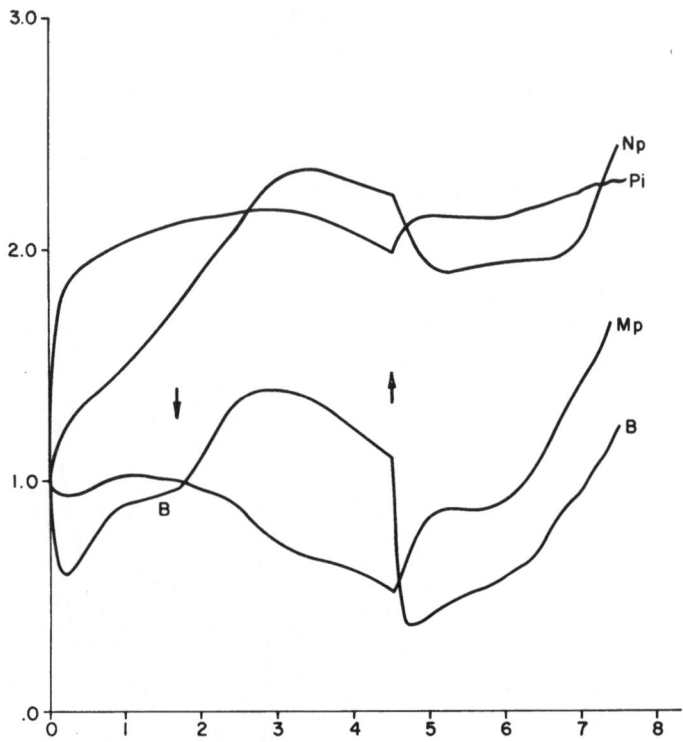

Fig. 76. Same experiment as in Fig. 75. Template N_p, messenger M_p, Pool P_i, and ribosome B are recorded during the experiment.

centrations of other functional entities were highly reduced. It is evident that the growth rate which establishes itself in the last end of the experiment is fairly comparable to normal growth. This can be seen in Fig. 77, and M^*, which represents normal messenger growth without injury, is compared with the formation of pool P_n. In summary, the model-system which goes through a temporary injury is capable of self-recovery when this injury is removed. However, immediately after termination of the injury some very drastic adjustments occur in organization. That such a system can establish its balance again is seen on the computer solution when observation times are further increased.

The question can be raised whether a cell with normal

functional characteristics can tolerate a partial genetic damage and still continue to grow. Such genetic damage might be carried through several generations, or it could be a permanently acquired characteristic. For example, a genetic error could be produced in the cell as a result of an injury. The question is how the growth is affected in subsequent generations. It has to be assumed, of course, in a cellular system that this type of damage does not interfere with DNA division, but merely reflects the fact that the affinity between RNA polymerase and the gene has been reduced. Of course, one has to assume that the genetic code has not been altered. One can speculate whether a cell which has a stable organization can reorganize itself after partial genetic injury. In order to elucidate this

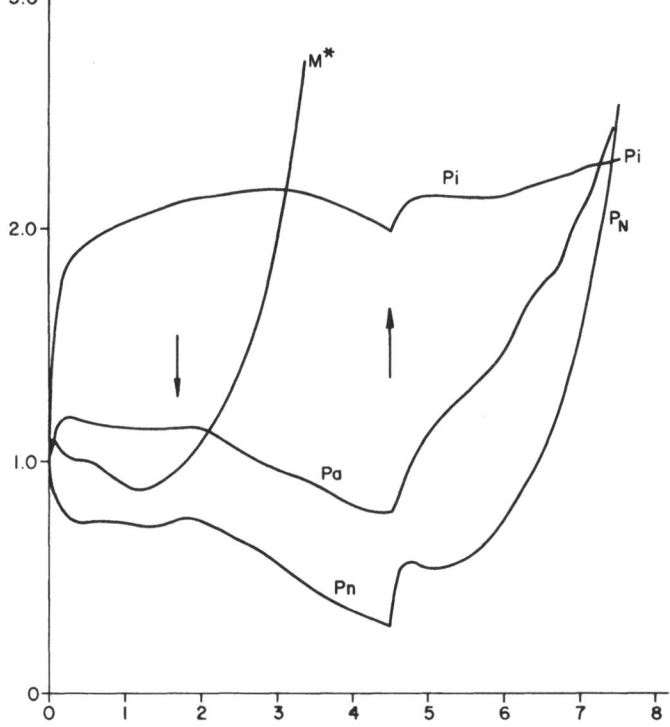

Fig. 77. Same experiment as in Fig. 75. Pools P_i, P_a, and P_n are recorded during experiment. M* indicates the normal growth of messenger M without genetic block.

problem, a simulation experiment was performed under conditions of 30% genetic damage. This genetic damage was introduced into the system by reducing k_6 at the time indicated by the arrow, in Fig. 78 and 79, and permanently maintained. Figure 78 shows that there is again an initial reduction of messenger M and template N concentration. Both pass through a minimum and then start to increase. Enzyme E is slowly reduced, but after a long delay will gradually start to grow. There is a continued growth of transport RNA C. It reaches a steady value, declines, and finally starts to increase again. Figure 79 shows that there is initially a rapid reduction of ribosome B, but there is shortly a rapid increase of ribosome

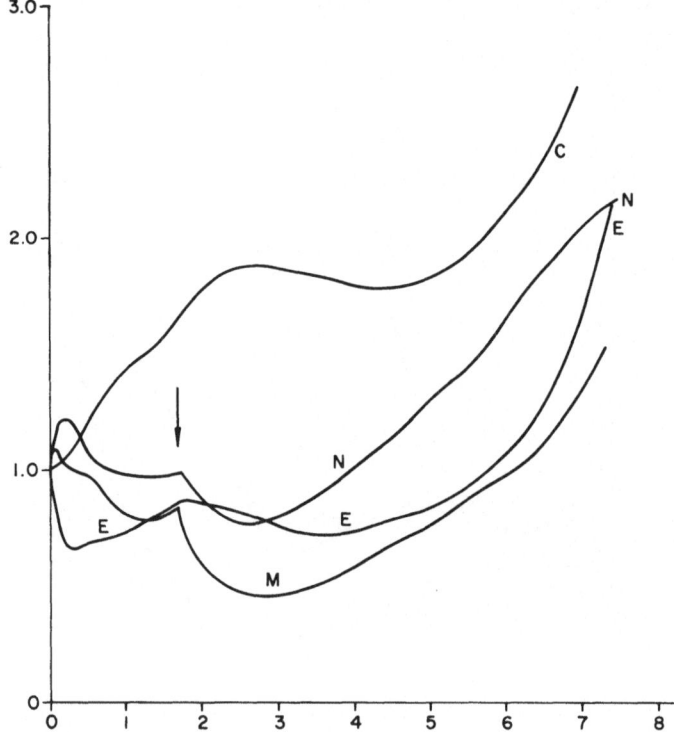

Fig. 78. The effect of partial genetic block at gene G_E level. The k_6 is reduced by 30% at the time indicated by the arrow and left in indefinitely. Messenger M, template N, enzyme E, and transport RNA C recorded during the experiment.

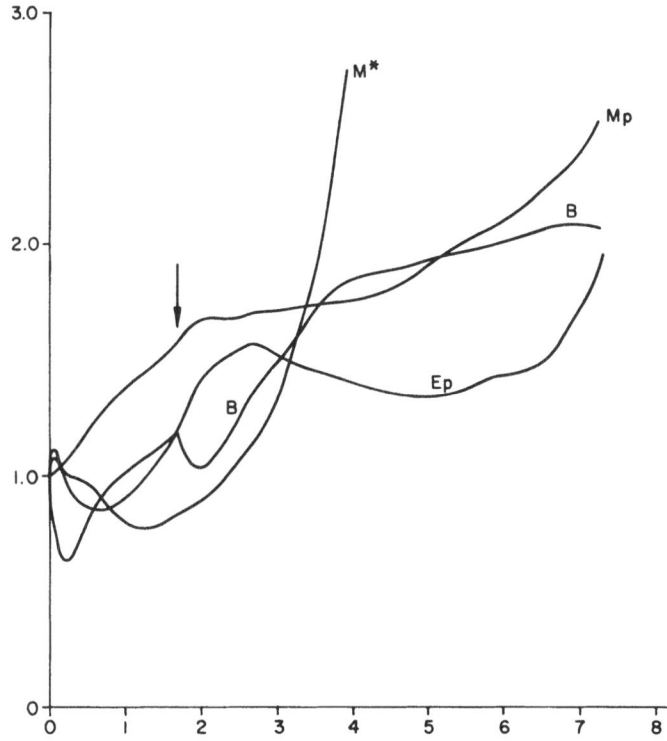

Fig. 79. Same experiment as in Fig. 78. Messenger M_p, enzyme E_p, and ribosome B are recorded during the experiment. Messenger M^* shows the normal growth without genetic block.

concentration, which finally levels off. Part of the initial ribosome transient arises from regulatory complex $[E_pB]$ formation, since E_p polymerase continues to grow when a genetic block is introduced at G_E level. However, E_p will subsequently start to decline, but finally will start to grow again when other elements increase. It is interesting to note that the rate of growth in conditions of genetic damage is considerably less when one compares it with the normal rate. Here messenger M^* represents the normal system, while M_p represents the system where a partial genetic block was introduced. Pools which are not shown here show that internal pool P_i does not, while pools P_n and P_a do, suffer initial transient when genetic block is introduced. There is a reduction in P_a and P_n concentration for a time, and then there is a gradual increase followed by a rapid growth. The final phase of rapid increase is always associated with an increase of all other functional entities in the system.

It is evident from these studies that the model-system is capable of compensating for its imbalance and will continue to grow, but does so at a reduced rate. However, growth rate reduction is not as drastic as one might expect. The principal reason for this is the phenomenon that the complex $[E_p P_n]$, which is essential for messenger M formation, will build itself up at a much higher level than in the normal state after the reduction of the k_6 value. Thus, the system automatically compensates for the injury, and as a result there is a partial increase in messenger M formation. Another mode of adjustment is the increase of transport RNA C concentration, which produces a higher concentration of $[CP_a]$ complex. As a result there is a partial increase in protein synthesis. These secondary compensations permit partial cancellation of some genetic damage. While the establishment of organized growth takes several generation times before all elements come into a smooth interrelationship with each other, it is quite evident that when genetic injury is not too deep, readjustment within the model-system takes place. However, we have not attempted here to find a critical level of injury where a readjustment does not take place. As we have seen before, 93% of injury was too drastic for the model-system to sustain, while here at 30% injury, the system is capable of reorganization and further growth.

Defects in Transport RNA Synthesis at Gene G_C Level.   Gene G_C activity can be controlled by varying rate constant k_4 (equation 4, Table II). It is evident from equation 6, Table III, that the synthesis of transport RNA is determined by the first term and decay by rate constant k_{15}. Others are regulatory or interaction type of terms in functional processes. In the following experiment the rate constant k_4 value was reduced by 93% at the time indicated by the arrow. This partially active gene G_C

was left permanently in the system. This arrangement in the
model-system simulates cellular growth under conditions in
which synthesis of transport RNA has been drastically reduced.
In order to carry out the analysis, equations in Tables II and III
and experimental data in Figs. 80–82 should be consulted.
The first result of the gene G_C inactivation is the rapid reduction of transport RNA C concentration. This is associated with
an increase of pool P_n. Concurrently with reduction of C there
is a reduction of [CP_a] complex concentration (not shown on
the graphs). The reduction of C concentration can be compared with the C* concentration, which represents normal
growth. It is evident that there is a tremendous reduction of

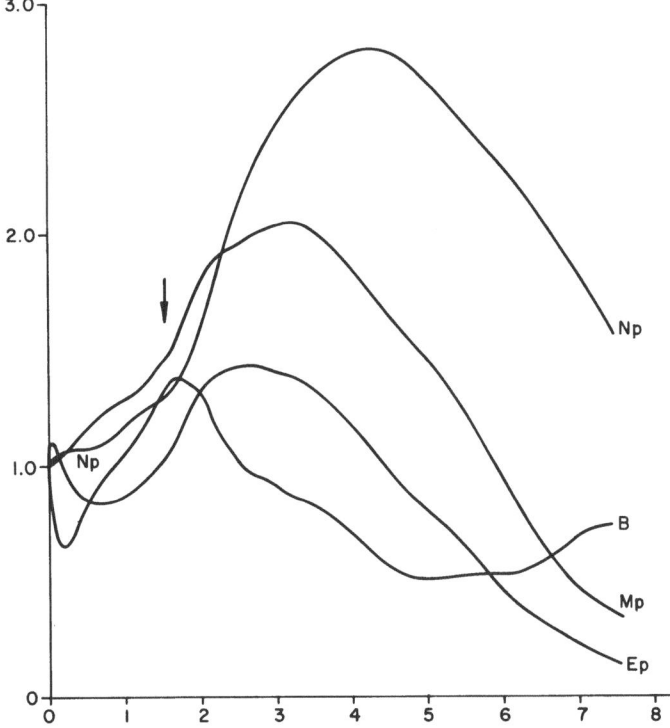

Fig. 80. Genetic block at gene G_C level. At time indicated by arrow, k_4 is reduced
93%. Template N, ribosome B, messenger M_p, and enzyme E_p recorded during the
experiment.

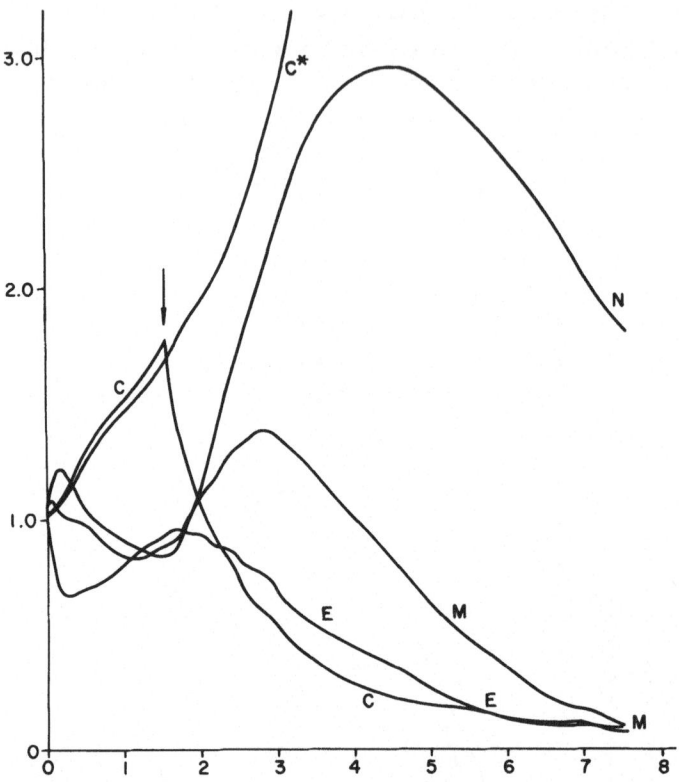

Fig. 81. Same experiment as in Fig. 80. Messenger M, template N, enzyme E, transport RNA C recorded during the experiment, C* indicates the normal growth without genetic block.

C concentration.   As a result of this $[CP_a]$ is small and there is a large increase of P_a (Fig. 82). Since there is a large increase of P_n, there is also a large increase of $[E_pP_n]$ complex. Consequently one expects an increase of messenger M and M_p formation. Such initial increase can be seen in Figs. 81 and 82.   There is also a very large initial increase in template N and N_p concentration, which is associated with a large reduction of ribosome B concentration. Enzyme E_p concentration also continues to increase, but enzyme E starts to decline initially.   All functional entities, which increase initially, reach a maximum value and subsequently start to decline.

Maximum concentrations of various entities are reached at different times. For example, P_a concentration grows for a long time, reaches a plateau, and then starts a steady decline. On the other hand, pool P_i immediately begins its gradual decline.

It is evident that enzyme E synthesis is very sensitive to the reduction of transport RNA synthesis. This is of course to be expected, since the $[CP_a]$ complex, which becomes small, is necessary for enzyme formation. Since in our system enzyme E is less stable than E_p, it is evident that the latter is less affected. The evidence is that protein synthesis in general is the first victim of gene G_C inactivation. Since most of the entities started to increase when gene inactivation took

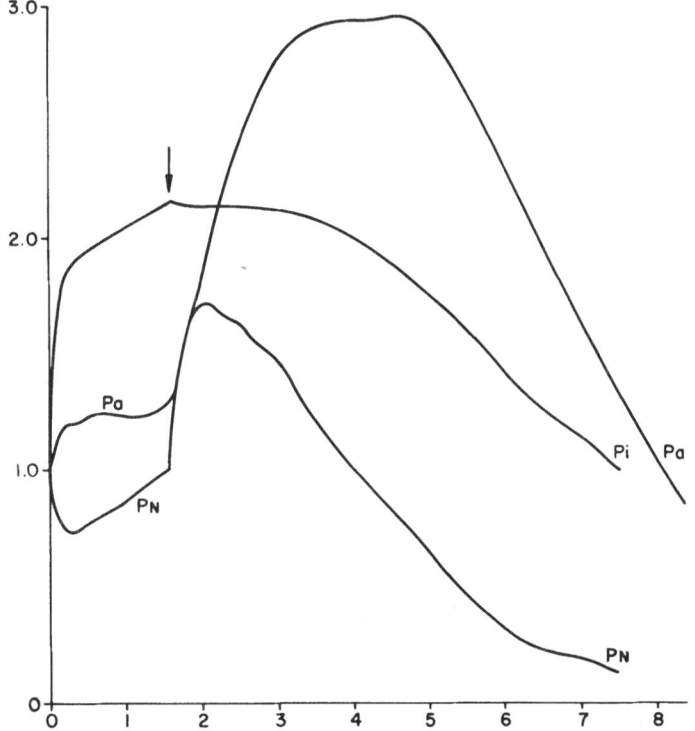

Fig. 82. Same experiment as in Fig. 80. Pools P_i, P_a, and P_n recorded during the experiment.

place, the question is: What are the limiting factors in this type of increase, and in what sequence will the functional elements start to decrease? The first functional entity to be affected is enzyme E. In the sequence, this is followed by ribosome B reduction, which follows closely the enzyme reduction. Pool P_n, after a rapid rise, starts to decline, followed by enzyme E_p, messenger M, and messenger M_p. All these entities are declining steadily while templates N and N_p still continue to grow. Finally, templates N and N_p also start to decline, while pool P_a is still growing. After a certain time interval (about 3 "generation times") all entities continue to decline. However, during the last phase of the experiment, ribosome concentration starts to increase again. There are many factors involved in such a phenomenon, such as, first of all, the leveling off the pool P_n, the release of B from regulatory complex $[E_pB]$, and the stability of B. However, observations on the computer with an extended time scale reveal that B reaches a secondary maximum and starts to decline again. Finally all entities approach zero level. Ribosomes, being a most stable entity in our system, are the last elements to disappear. It appears that 93% inactivation of gene G_C activity is sufficient to disorganize completely the functional system.

It is of great interest to find out whether the system is at a certain time capable of recovering from a genetic injury, which is leading potentially to a nonfunctional state, and what factors will affect recovery. The first question is, in the framework of given system, how long can the system sustain a genetic injury and still be capable of recovery In order to perform recovery experiments, the system was slightly modified so that some of the transients, especially the templates, would not grow too large. Secondly, in order to reduce the time during which the growth transients resolve, we introduced a complete genetic block at the time indicated by arrow. This is essential, since observation time is limited on the computer. In the first experiment, genetic injury was maintained for a short enough time so that the system was capable of recovery. Subsequently, we shall study a case in which the duration of injury is too long for the system to recover. Figures 83, 84, and 85 show the experiment in which gene G_C injury is temporary and the recovery process takes place. The first event when a complete gene block is introduced is rather similar to the experiment described previously. However, the events following the gene reactivation are of interest to follow in detail. Here the rate constant k_4 was reduced to zero at a certain time, and after an interval it was restored to its original value (Fig. 85). It is of interest to note that during the complete genetic block all components of the system started to decline and become zero after 8 "generation times." In the experiments with a temporary block the

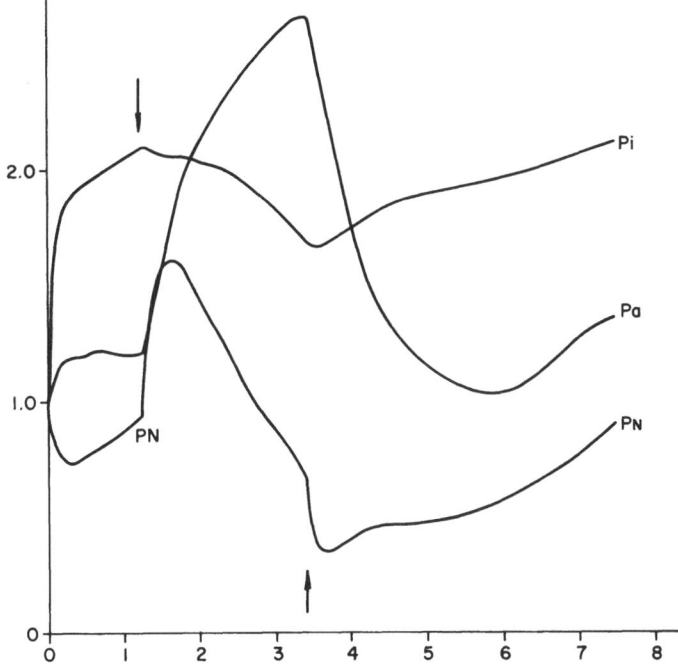

Fig. 83. The effect of complete genetic block on gene G level. At the time indicated by first arrow, k_4 is made zero.. At time indicated by second arrow, the genetic block is removed. Duration of genetic block is such that cell is capable of recovery. Pools P_i, P_a, and P_n are recorded.

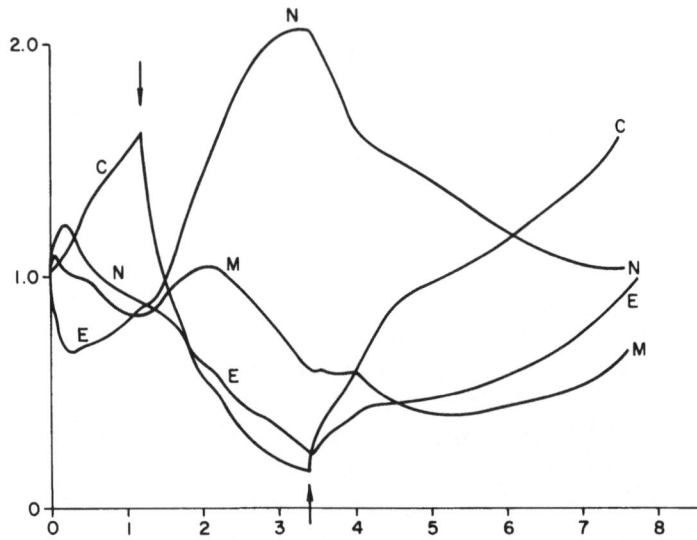

Fig. 84. Same experiment as in Fig. 83. Template N, messenger M, enzyme E, and transport RNA C are recorded.

recordings were made during 5 "generation times." The first event taking place when the gene is reactivated is the increase of transport RNA C concentration, which is shown in Fig. 84. This is associated with the decrease of pool P_n and pool P_a while pool P_i starts gradually to increase. Practically at the same time an increase of enzyme E and ribosome B occurs. In contrast, M_p and E_p both continue to decrease. Both templates N and N_p were near the maximum concentration when k_4 was restored to the original value, and both started immediately to decline. Template N declines initially very rapidly, but after a while reduction proceeds at a slower rate. Messenger M shows initially practically no change, but subsequently there is a decrease of M concentration. After passing a minimum, M gradually starts to increase again. The recovery of the system takes place rather unevenly, and various functional entities which were reduced start to increase at different times. The first to increase is enzyme E_p, then messengers M_p and M, followed by pool P_a. Finally, at the end of generation time, N_p starts to increase while N reaches a

constant level.   N finally also starts to increase, but this can
be seen only when the observation time is extended on the
computer.

This experiment reveals that introduction of the genetic
block really disorganized the system.   However, removal of
the block initiated a sequence of events which permitted the
system to recover.   The recovery is not uniform, however,
and it will take numerous generations to stabilize the system.
An important conclusion which can be drawn from this simulation experiment is also that a cell may be capable of self-
recovery from a prolonged genetic injury.   In these experiments duration of the genetic block was slightly longer than a
generation time.   This raises the question as to what is the
minimum duration of genetic block which would prohibit
recovery of the system.   In order to explore this the experiment was performed with the same model-system, but the
growth rate was slightly reduced by increasing k_{30} and k_{17} and

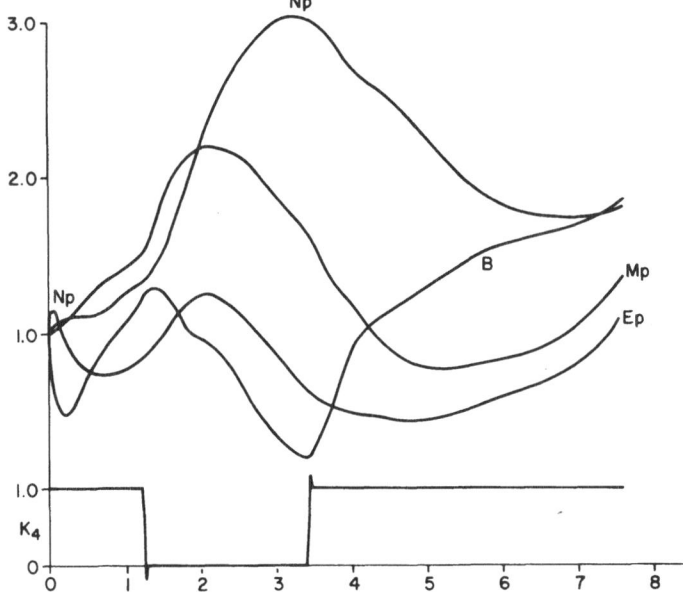

Fig. 85. Same experiment as in Fig. 83.   Template N_p, ribosome B, messenger
M_p, enzyme E_p are recorded.

by decreasing k_{18}. This was necessary in order to avoid computer overloading in transient phases of the experiments. Data in Figs. 86–88 reveal that, starting from initial conditions, the cell grows more slowly than in previous experiments (Figs. 83–85). In the absence of the genetic block, this system starts to grow slowly and continues to do so indefinitely. Rate constant k_4 was reduced to zero at the same time as previously, but duration of the zero period was markedly increased. In these conditions, after removal of the genetic block, the system was not able to recover and continued to decline. At first, we should like to review the sequence of events (Figs. 83–85) when the system recovered. After removal of the genetic block, there was an immediate rise of transport RNA C, enzyme E, ribosome B, and pool P_i, together

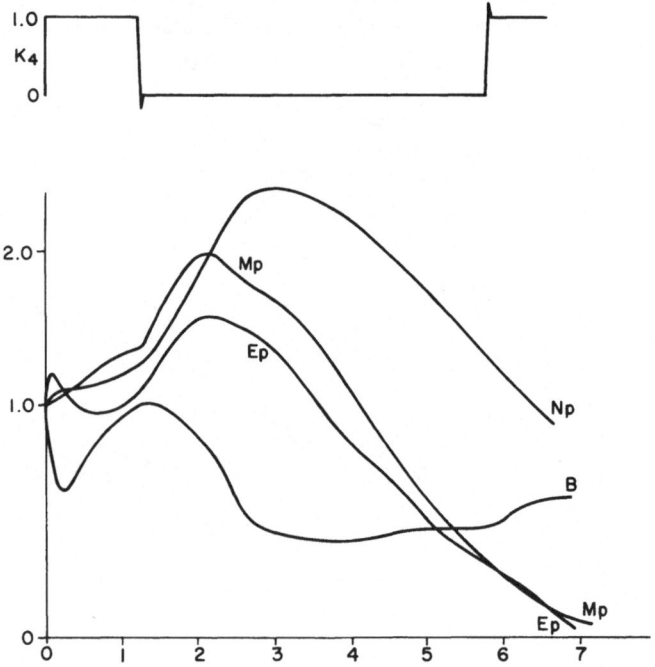

Fig. 86. This is the same experiment as in Fig. 83, except that the duration of block is longer. System does not recover from such injury. Template N_p, messenger M_p, enzyme E_p, and ribosome B are recorded during the experiment.

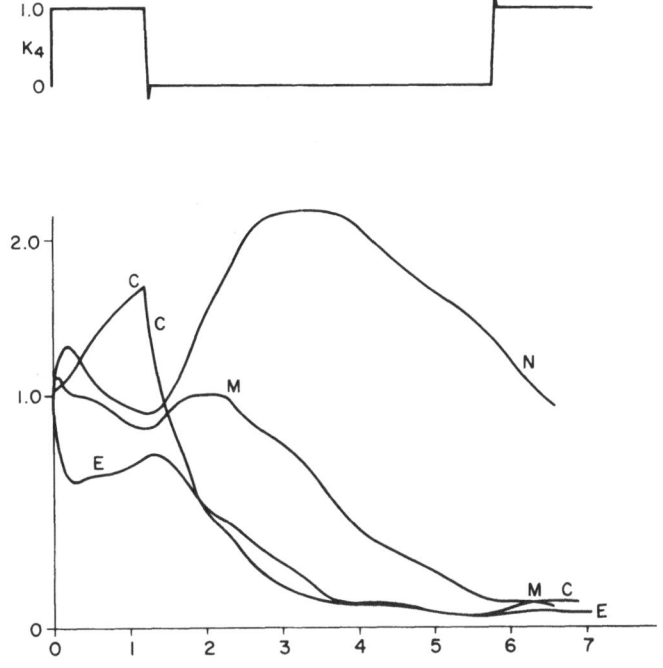

Fig. 87. Same experiment as in Fig. 86. Template N, messenger M, enzyme E, and transport RNA C are recorded.

with an immediate decline of pool P_n. In the system (Figs. 86–88) which does not recover after removal of the block there is a small increase of ribosome B, but the rise is very small compared with the previous experiments which recovered from the injury. Also, transport RNA C has a very small increase in concentration, but it is trivial and C does not continue to grow. Enzyme E has practically no increase, and the concentration is maintained at a low level. Pool P_i remains constant for a while, then starts to decline. There is a small reduction of P_n. The kinetic behavior of the system indicates that synthesis initiated after the removal of the genetic block is extremely small, and this synthesis is not sufficient to make the system again operational. It is instructive to observe that the secondary, delayed increase of several functional entities, such as messenger M_p, template N_p, and polymerase E_p, is

completely absent here, while these were observed previously
(Figs. 83–85). Furthermore, none of the pools show increase
(Fig. 88) after restoring k_4 to the normal level. It appears
that the system has a very small initial increase in concentration of some functional entities after removal of the genetic
block, but this is not sufficient for the reorganization of a
properly balanced synthesis, and the system will decay out.
Some further experiments were carried out on the computer
to determine critical duration of the genetic block. Experiments revealed that if the blocking time is slightly reduced,
the cell is indeed capable of self-recovery. Consequently, the
experiment reported in Figs. 86–88 represents rather the
borderline conditions, where the system loses its ability to
recover. When blocking time is further increased, the system
will decay rapidly. It appears that there is a critical time limit
from which a given system can be exposed to a particular type
of genetic block.

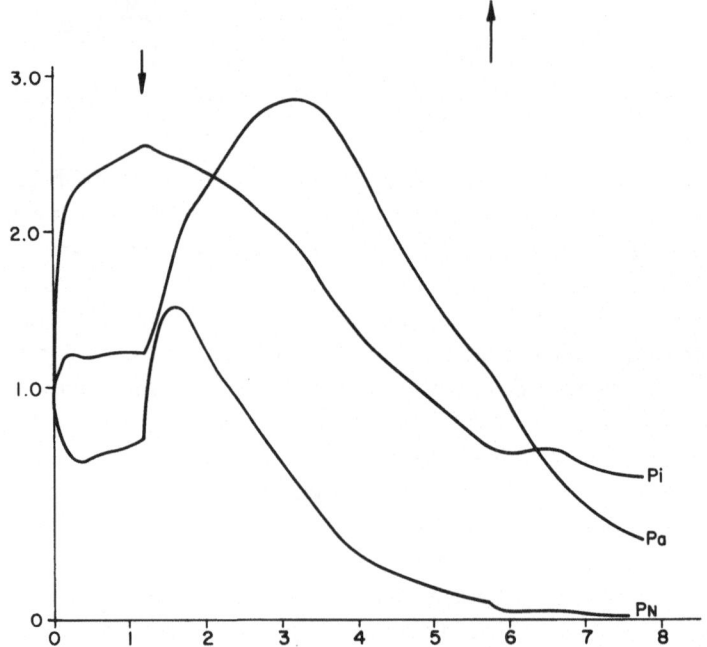

Fig. 88. Same experiment as in Fig. 86. Pools P_i, P_a, and P_n are recorded.

This simulation experiment represents a very important concept in cellular injury analysis, because it clearly reveals the significance of the time factor in the process of recovery. However, if the genetic block is not complete, then different time relationships appear and only systematic study can give useful information in this area. These simulation experiments on the model-system suggest some considerations for cellular systems. By an application of a temporary genetic block, the model-system is terminated and the question can be asked whether there is a requirement for permanent genetic injury to kill the cells. It appears that perhaps a certain length of the blocking time is sufficient during cellular growth to inactivate the cell. If this were the case, then differential killing of different cell types could be possible simply on a time basis. If we assume that a long-lasting genetic block can kill off the cell, then the question can be raised as to whether all cells in a population can be killed off with the same critical block lasting a definite time. This, of course, seems possible if one assumes that all cells behave equally. This assumption, of course, is contrary to practical experience, since there are variations in various biological properties among individual cells. This brings up the point of whether the properties of various functional entities will affect the critical blocking time in the model-system and how much variation is necessary. This problem, of course, would require a systematic study, which is not done here, but we shall present a few examples to demonstrate that the model-system can be used to analyze such a problem. Furthermore, we shall show that stability variation in the values of individual functional entities is the determining factor in determining the viability of the system.

The effect of enzyme E stability was studied at first. Rate constant k_{16} characterizes this property in the system. In various experiments k_{16} was varied within a certain range and observations were carried out. In order to study this effect more clearly, the observation time on the computer was extended so that kinetic behavior of entities could be observed during 7 "generation times." In previous experiments, shown in Figs. 86–88, the rate constant k_{16} value was 0.0170. At this rate constant value, the system was not capable of recovery. When k_{16} was varied on the computer, visual observations revealed that a borderline value for k_{16} was 0.0143. At this level the system started to recover, but the recovery was extremely slow and not suitable for recording. Recovery of the system was recorded at a k_{16} value of 0.0107, and the data is presented in Figs. 89 and 90. It should be noted that changing the rate constant k_{16} from a normal value of 0.0170 to the value 0.0107 introduces some changes in the normal growth rate. This can be observed when this experiment is compared with the previous one at the time interval before the genetic block is introduced into the system. The increase in the stability of enzyme E will also increase the growth rate of functional

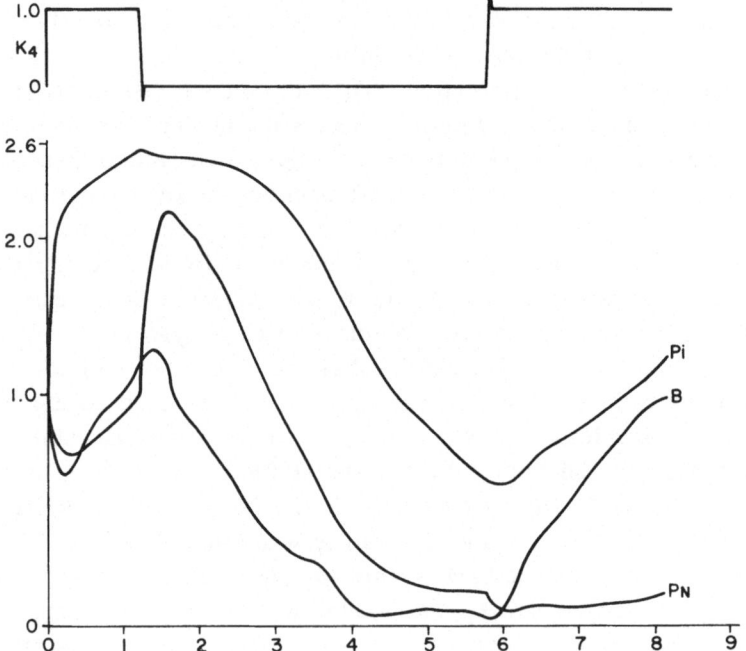

Fig. 89. Same experiment as in Fig. 86, except that protein stability is increased. Rate constant k_{16} is reduced from 0.01702 to 0.0107. This increased enzyme E stability permits recovery of the system. Pools P_i, P_n, and ribosome B recorded.

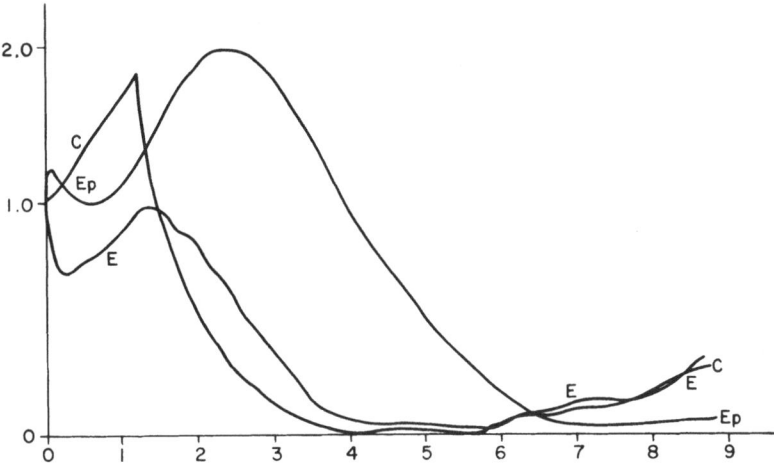

Fig. 90. Same experiment as in Fig. 89. Enzymes E and E_p and transport RNA C
are recorded.

entities as a function of time. When the genetic block is introduced at the same time as previously by reducing k_4 to zero,
it is evident that general features of the response of the system
are similar to previously described experiments in Figs. 86–88.
It is evident from these experiments that some values of
functional entities become extremely low during the genetic
block. For example, transport RNA C concentration will become extremely small. Enzyme E concentration also declines
steadily. When the genetic block is removed, there is an immediate increase of C and E, and a reduction of P_n. Furthermore, there is a slightly delayed, but pronounced increase in
ribosome B and internal pool P_i. This response is similar
to the recovery which was analyzed in the previous experiments
(Figs. 83–85). There the response was much sharper, but the
fundamental phenomenon is the same, so that certain basic

functional entities should be able to grow in order to restore the functionality of the system.

In order to determine some other factors which are important in the recovery of the system, visual obserations were carried out on the computer. These revealed that the stability of templates and messengers strongly affects the ability of the system to recover. These experiments emphasize the all-important aspects of protein synthesis in the process of recovery. Therefore, stability of elements which participate in protein synthesis are important in determining the stability of the system and its ability to recover. This ability to recover depends not only on the stability of the functional element, but also on the rate at which individual functional elements are formed. Consequently, complex variations in reaction patterns provide multiple possibilities for the viability pattern of the system.

Genetic Injury at Gene G_B Level.   The blocking of ribosomal gene G_B will produce essentially the same results as has been observed in genes G_C and G_E, except that the secondary responses in growth adjustment are very drastic. Blocking of the gene G_B by making $k_3 = 0$ will free a large part of pool P_n. Messengers and transport RNA are rapidly built up, and a very distorted organization pattern results. If the block is maintained for a sufficiently long time, the system will decay until it ceases to exist, but the process is extremely nonuniform as far as balanced organization is concerned. There are excessive concentrations of some functional entities, but very low concentrations in others. One can project that cellular death in these conditions would represent morphological changes of an extreme kind. Since the process of injury at gene G_B level is so disorderly, it is not considered worthwhile to carry out a detailed analysis of injury and recovery mechanisms.

Injury at Gene G_P Level .   Because of the complexity of the model-system, it was impossible to introduce more than one functional polymerase into the system. Since gene G_P controls total enzyme synthesis, it was of interest to study the viability of the system when injury occurs at that site. Rate constant

k_5 was varied for that purpose. Experiments on the computer reveal that blocking of gene G_P will also produce drastic distortions in the system. No attempt was made to study injury and recovery mechanisms further at G_P level.

d. Nongenetic Injury and Death

General Comments. Inactivation of various functional entities which are nongenetic in character is the subject of this study. Since genes comprise but a small fraction of all cellular elements, it is evident that nongenetic injury plays in all prob-. ability a much larger role in nature than does genetic injury. Genes are organized principally as chromosomes, so that genetic organization is thus confined mostly to the nucleus. Extra-chromosomal genes are physically dispersed, but they operate as distinct operational units. This is in contrast to the distribution of enzymes, which are spread throughout the intracellular volume and membrane structures. The enzyme is a building block of the cellular system; it is a common building block, like a brick in the house, which operates at every possible locus in the cell. This type of entity does not represent any specific physical, well-confined target. It is evident that the target concept for nongenetic injury is rather meaningless, although it does have meaning in genetic injury, where location of the functional unit is well defined. In order to analyze injury mechanisms via the model-system, it appears that a more specific understanding of cellular processes can be obtained when a systematic study is carried out on the level of individual functional entities. Specific entities are to be inactivated individually, and not simultaneously with other components. In simulation of cellular injury, the inactivation of functional entities of the system is carried out so that an inactivating agent interacts with a particular functional entity. The interaction product is not functionally active and interaction is irreversible.

We are simulating here cellular viability and recovery. Furthermore, we attempt to elucidate the conditions which are essential for cellular recovery. Model-system study simulates the experiments where growing cells in the medium are exposed to an external agent producing damage during growth. In order to carry out these experiments on the computer, a specific rate constant is increased or decreased to a desired value at certain time and maintained at that level for a time. Then, after an interval, the system will be brought back to the normal state. This could simulate, for example, the conditions that at a certain time interval all enzymes are subject to rapid inactivation. Assume that entity A is normally decomposed by a process which is characterized by a rate constant k_x. At a time t_1 a process of inactivation occurs (rate constant k_x^1), which is terminated at time t_2. The end product is the same for both processes.

The following reactions are indicated:

$$\text{Time Period:} \quad 1) \quad 0 \text{ to } t_1 \quad \overset{o}{A} = k_x A$$

$$2) \quad t_1 \text{ to } t_2 \quad \overset{o}{A} = (k_x + k_x^1)\, A$$

$$3) \quad t_2 \text{ to } t_3 \quad \overset{o}{A} = k_x A$$

During the period t_1 to t_2 inactivation of A takes place.

First, we shall consider inactivation of enzyme E. Since all enzymes are proteins, one could consider that inactivation of enzyme E and polymerase E_p could be carried out at the same time. However, it seems more instructive to study the effect of inactivation of those enzymes separately because they have distinctly different functional characteristics. Enzyme E is essentially a component, building up the intracellular pools and intermediates, while enzyme E_p polymerase is essential for RNA formation. By dissecting out individual elements and studying their effect on injury processes, a much clearer understanding of the complex events will be obtained. Consequently, we shall analyze the inactivation of those functional entities separately.

Activation of Enzyme E. Enzyme E stability is characterized by the rate constant k_{16}, and in the normal state k_{16} has a value of unity. At a certain time interval, the enzyme is inactivated, a state which can be simulated by an increase of k_{16}. This

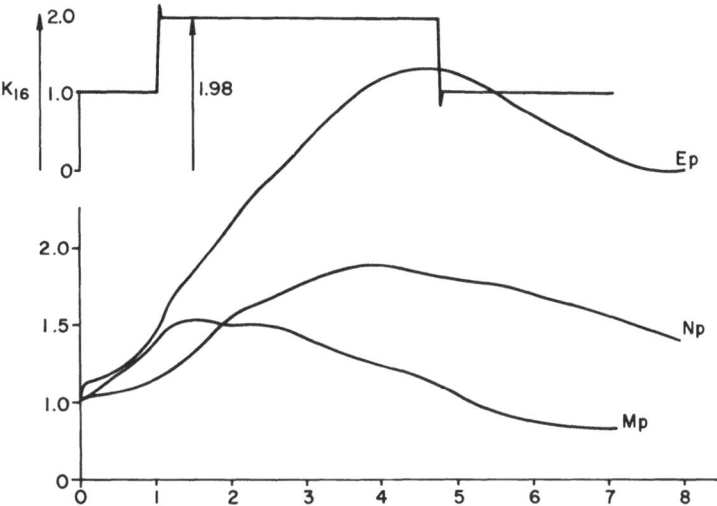

Fig. 91. The effect of temporary inactivation of enzyme E. Rate constant k_{16} changed from value of unity to value 1.98, as indicated in the graph. System does not recover from this injury. Enzyme E_p, messenger M_p, and template N_p recorded.

experiment is presented in Figs. 91 and 92, where it is evident that at a certain time k_{16} was increased from the value 1.00 to 1.98 and after a certain time interval k_{16} again returned to unity. When k_{16} is increased, there is immediately a sharp drop of enzyme E concentration, subsequently followed by a steady decline. During the time when k_{16} is large, E is steadily declining; while synthesis of other functional entities continues, their concentration increases to a maximum value and then will start to decrease. Observation time in these experiments is 6 "generation times," which permits us to follow all sequential events. The first components which start to decline are transport RNA C and messenger M_p. These are followed by messenger M and template N. The last entities to start to decline are template N_p and polymerase E_p. Since the E_p polymerase system is competitive with the enzyme E system, it is clear that a reduction of the E value will benefit the E_p polymerase system. At a time when k_{16} returns to the normal state (which is equivalent to that external agent being removed from the system),

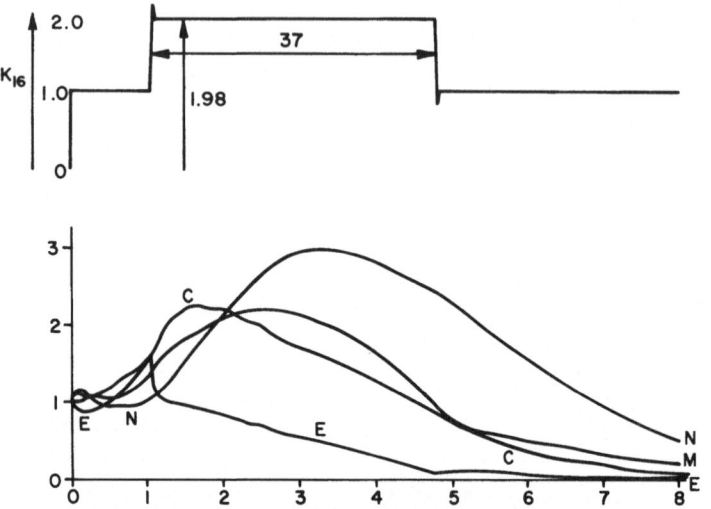

Fig. 92. Same experiment as in Fig. 91. Template N, messenger M, enzyme E, and transport RNA C recorded.

enzyme E has reached practically zero level. This is contrasted with enzyme E_p, which at the time of the k_{16} return to unity is at an approximately fivefold concentration above the normal state. Templates N and N_p also reveal relatively high concentration values, while transport RNA C and messengers M and M_p are smaller, but their level is still substantial. It is remarkable that the reintroduction of k_{16} has a relatively minor effect on the course of events. All entities continue to decline. This system does not exhibit any symptoms of recovery to the normal state. Extended time scale observations on the computer show that this system decays completely. It is evident that enzyme E concentration, at the time when k_{16} returns to unity, has a slight transient, and that there is a region where the enzyme concentration is kept constant for a while, although it subsequently starts to decline steadily. It seems that the system is not capable of reorganizing itself to initiate further growth. Other functional entities, which are not represented in the figures, reveal the same general characteristics.

In summary, this simulation experiment suggests that when an enzyme system, excluding polymerase, is inactivated during cellular growth, and when the inactivating agent is removed, the cell may not be capable of recovery. However, recovery may depend on the exposure time. To explore the time effect, the interval of k_{16} application was varied. Experiments show that when the time is reduced, the system is again capable of recovery. The question is whether the property of some entity could affect the recovery. For example, can the system recover when it contains a more stable protein? For this purpose, experiments were carried out, and conditions were determined at what level of protein stability the system could recover. Figures 93 and 94 show the experiment where the decrease of k_{16} from the value 1.98 to 1.92 was sufficient to produce the recovery of the system. It is of interest to note that a 3% increase in enzyme stability was sufficient for the system's recovery. The sequence of events leading to recovery is of great importance. Of principal interest is the kinetic behavior of enzyme E. Immediately

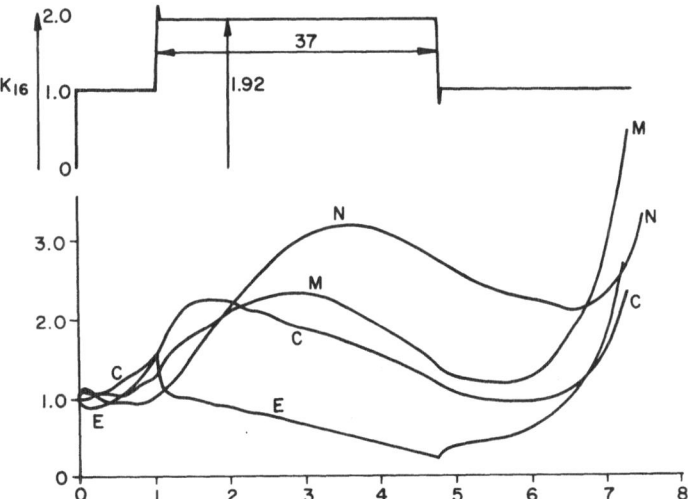

Fig. 93. The effect of inactivation of enzyme E. Same as in experiment Fig. 91, execpt that k_{16} is increased from value 1.0 to 1.92 (not 1.98, as in previous experiment). Recovery of system occurs in this condition. Template N, messenger M, transport RNA C, and enzyme E are recorded.

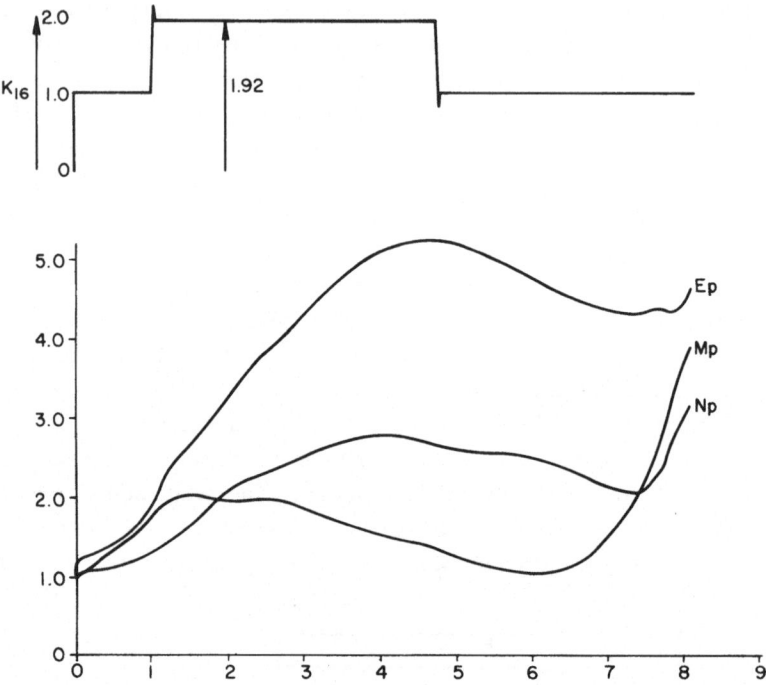

Fig. 94. Same experiment as in Fig. 93. Enzyme E_p, messenger M_p, and template N_p are recorded.

after the return of k_{16} to a normal value of unity, enzyme E concentration starts to increase. The increase is small, but distinct. Subsequently, E concentration continues to increase, first slowly and then rapidly. (This is in contrast to behavior of E, as exhibited in Fig. 92.) It is noteworthy that E is increasing while other entities are still declining. The next component to increase is messenger M. Shortly after the increase of M concentration, transport RNA C also starts to increase. Finally, messenger M_p and template N will start to increase, and subsequently all entities start to grow rapidly. The sequence of events which leads to new synthesis is the formation of messenger M, transport RNA C, and template N. The initial rapid increase of enzyme E seems to be an all-important event. It appears that in these experiments the regulatory compensation and synthetic balance were properly adjusted for initiation of progressive growth. Conditions for the initiation of the recovery process are indeed complex, since there is a multiple set of interactions. It is evident that the polymerase E_p system, including M_p and N_p, was never prohibitive for cellular recovery. When compared with the experiments in Fig. 92, it is evident that in order for systems to start to grow, it is essential that enzyme E and complementary synthetic components M and N all start to increase. This seems to be a prerequisite for cellular recovery. Figure 92 shows that there was indeed a small increase in enzyme E concentration when k_{16} was restored to unity, but there was never an increase in messenger M, template N, and transport RNA C concentration. It is apparent that a small increase in protein stability was sufficient to produce a viable system.

These experiments suggest that in a mass population where there are variations in enzyme stability among individual cells, cellular survival is variable. Furthermore, the variability of various functional entities is instrumental in contributing to the variability of cellular survival.

Inactivation of Enzyme E_P. Enzyme E_p inactivation can be carried out on the basis of the same premise which was presented previously for enzyme E. Instead of k_{16} here, the rate constant k_{17} (equation 17, Table II) value would be suddenly increased, maintained so for a certain time, and returned to normal level. This means that during the time when the increased rate constant is operational, inactivation of enzyme E_p takes place. Figures 95 and 96 show the experiment where k_{17} is increased from its normal value of 1 to a value of 4.8. After a certain time interval, k_{17} again acquires the value of 1. The first effect of the increase of k_{17} is the rapid, initial reduction of E_p polymerase concentration, which declines until it reaches a minimum. Then E_p starts at first to grow gradually, then more rapidly, passes through a plateau region, and finally establishes itself on a course of slow, steady growth. Messenger M_p suffers a change in growth rate, but continues

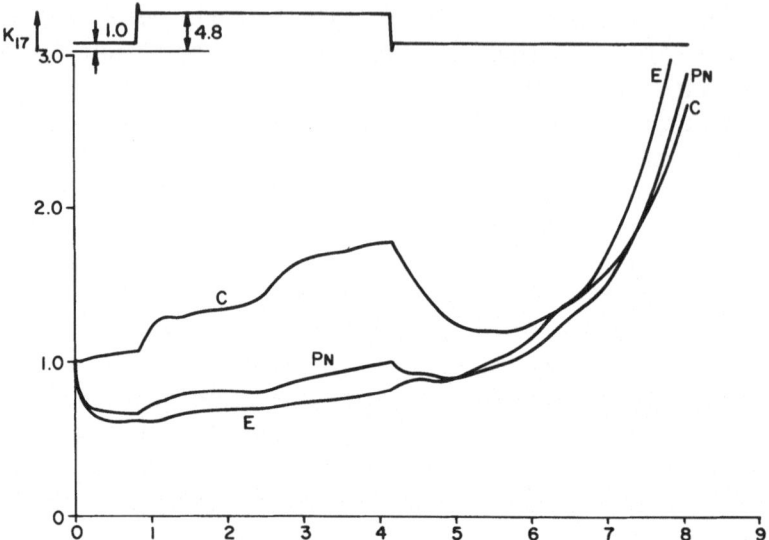

Fig. 95. The effect of enzyme E_p inactivation. Rate constant k_{17} increased from value of 1.0 to 4.8, and after certain time interval k_{17} is reduced back to value of 1.0. System recovers from this type of injury. Enzyme E, pool P_n, and transport RNA C are recorded.

to grow steadily afterward. Template N_p concentration proceeds straightforwardly without any changes when k_{17} is increased. Transport RNA C exhibits an immediate increase when k_{17} is increased. This transient is mostly caused by the liberation of C from the complex $[E_p C]$. This is a regulatory adjustment following the reduction of E_p. After this initial transient has passed, there is, for a while, a steady growth of C. Then there is a slow growth transient followed again by steady growth. Pool P_n and enzyme E increase slightly, and both continue to grow rather slowly. As a matter of fact, the increase of k_{17} to 4.84 will actually produce a slowly growing system. When k_{17} is reduced to a normal value of 1, there is immediately a rapid increase in E_p concentration. After a delay, this is followed by a slow increase of messenger M_p and enzyme E. There is a rapid reduction of transport RNA C, caused again by regulatory adjustment. After passing through

a minimum, C will start to grow again. After some time interval, all entities begin to grow rapidly.

When rate constant k_{17} is drastically increased and maintained at that level, a nonviable system results. It is of interest to see how the system, which is proceeding into complete
decay, is able to reorganize again when the k_{17} block is released. Figures 97 and 98 illustrate the experiment in which
rate constant k_{17} was increased from 1 to 11. This k_{17} value
was maintained for the same time period as in the previous
experiment (Figs. 95 and 96). It is evident that an increase
of k_{17} produces a rapid initial reduction of E_p concentration.
After passing through a minimum, E_p starts slowly to increase,
reaches a slight maximum, and subsequently will start to decrease again. The messenger M_p grows for a while, reaches
a maximum, and then starts to decline. Template N_p continues to grow and finally reaches a maximum at a relatively
high concentration value. Enzyme E has a small initial
transient, reaches a slight maximum, followed by a steady

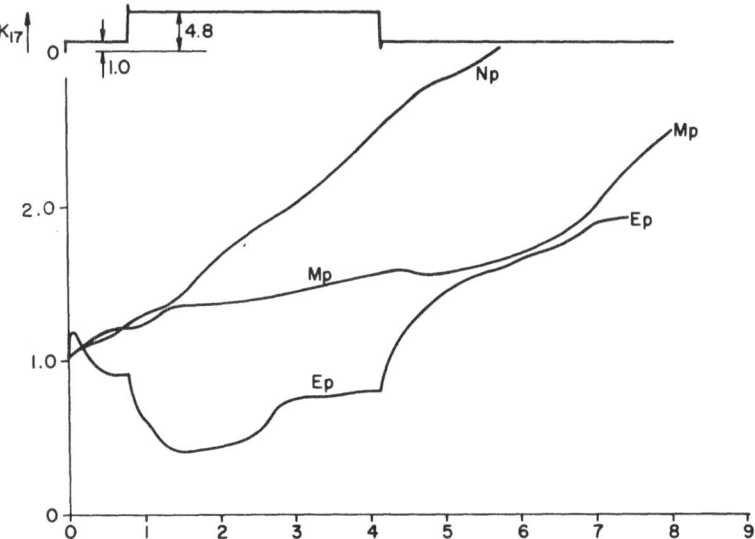

Fig. 96. Same experiment as in Fig. 95. Template N_p, messenger M_p, and enzyme
E_p are recorded.

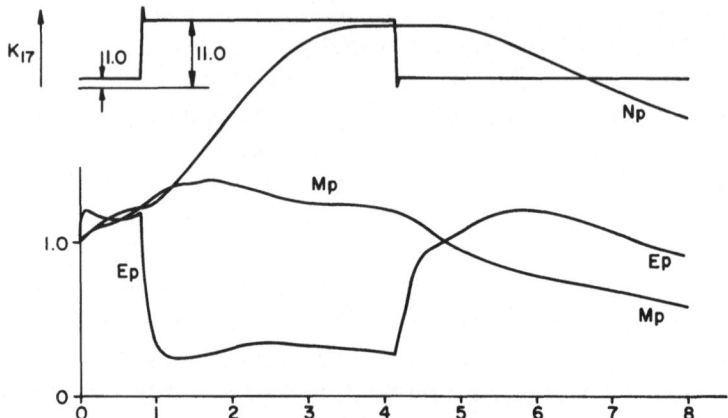

Fig. 97. Same experiment as in Fig. 95, except k_{17} value is increased from value 1.0 to 11. System becomes nonviable after this type of injury. Template N_p, messenger M_p, and enzyme E_p are recorded.

decline. Similarly, pool P_n has a rather rapid initial increase, reaches a plateau, and subsequently declines steadily. Transport RNA C increases rapidly in two stages, reaches a maximum, and starts to decline. All elements of the system which are not recorded here have been observed on the computer to show different kinetic characteristics, but all decline after

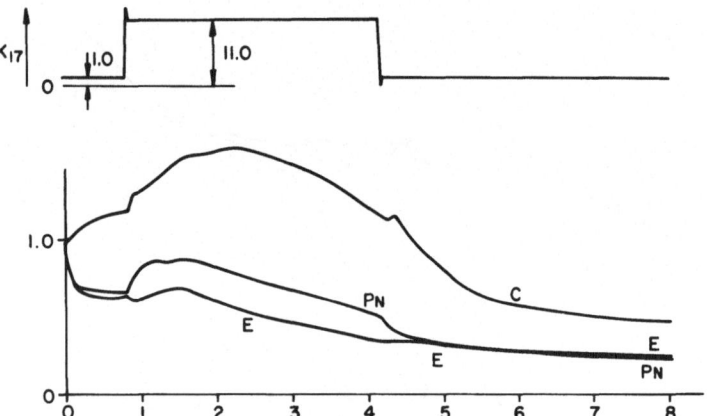

Fig. 98. Same experiment as in Fig. 97. Enzyme E, pool P_n, and transport RNA C recorded.

passing through a maximum, and if the k_{17} value is maintained at this level, the system will gradually become nonfunctional. In order to restore the functional system, k_{17} is brought back to the normal value of unity. However, it is evident that this system is not capable of recovery. This seems surprising, since E_p polymerase rises very rapidly. This is followed by a phase of less rapid growth until the maximum is reached, and then E_p decreases slowly. Messenger M_p, in contrast, continues to decline, even when the rate constant k_{17} is increased, but with a greater rate of decline. There is no phase of recovery, as can be seen in Fig. 96, where M_p concentration is fairly constant for a while and then starts to increase. Template N_p and enzyme E stay constant for a while, and subsequently start to decline steadily. Pool P_n is reduced slightly in the beginning, and then a steady decline occurs here also. Transport RNA C, after a small initial transient, starts to decay steadily. It is of interest to note that the concentration of enzyme E_p reached a rather high value when k_{17} was reduced, but this seems to have had relatively no effect on the overall growth of other functional entities. It appears that this system does not even attempt to make a recovery, since enzyme E at no time shows any increase.

Experiments presented so far reveal that drastic changes in the activity of enzyme E_p polymerase yield a nonviable system. It is of interest to consider whether it is possible to reduce E_p activity permanently to a level where the system is not capable of growth, but is able to maintain its functional activity at a constant level. Such an experiment would be significant for simulating two types of cellular processes. First, one could demonstrate that an external agent which interacts with the polymerase type of enzymes could suppress cellular growth while not affecting at the same time cellular viability. Secondly, an internal growth regulatory mechanism could be operational at E_p activity level. Such a condition could result when, for example, some specific hormone or substrate interacts with E_p and thereby reduces its enzymatic activity. Both phenomena are of great interest from the point of view of

cellular growth control via external and internal agents. Consequently, it was decided to study the model-system behavior
further in conditions where rate constant k_{17} is increased, thus
inactivating a large portion of E_p activity. The process could
be formulated as follows:

$$1. \quad E_p + A \xrightarrow{k^1} E_p^1 \quad \text{(inactive)}$$

$$2. \quad E_p^1 \xrightarrow{k^{11}} P_i$$

A represents an external or internal agent. In our experiments rate constants k^1 and k^{11} would be incorporated into a
single-step reaction at the level of k_{17} (equations 17, Table II).

In our previous studies (Figs. 95–96), it was evident that
some components of the system, especially N_p, would become
too large and exceed the operational limits of the computer.
Therefore, it was essential that the growth rate be sufficiently
reduced to avoid computer overloading during the long observation times (equivalent to about 7 normal "generation times"). For

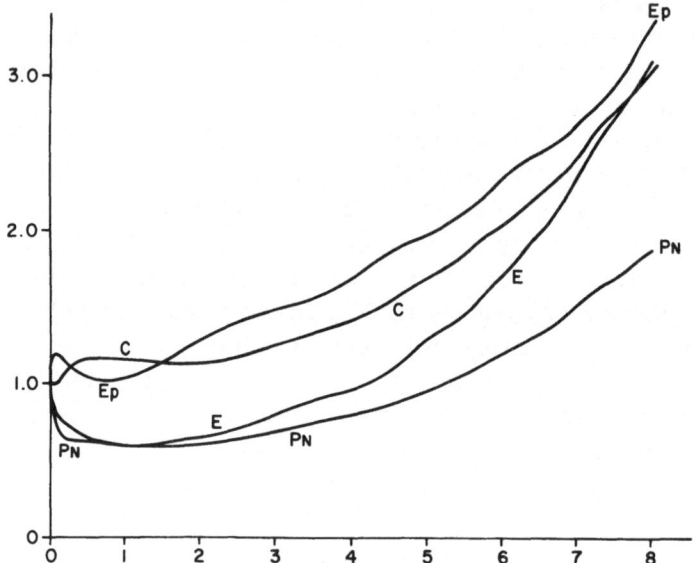

Fig. 99. Control of growth at the level of E_p polymerase. This is a normal, slowly
growing system and serves as a control for experiment in Figs. 100, 101, and 102.

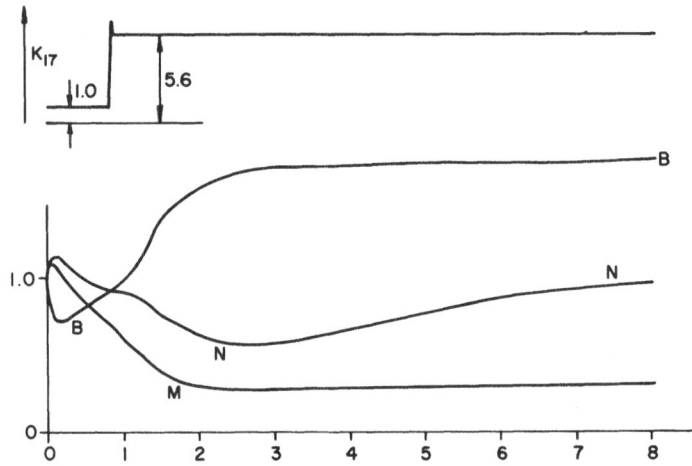

Fig. 100. Control of cell growth E_p polymerase activity level. Increase in k_{17} from value of 1.0 to 5.6. Under these conditions system does not grow.

this purpose, some rate constants were slightly changed, and external pool P_e values were reduced by 66%. Normal growth for such a system is very slow, and it is possible to follow the growth of individual entities for extended periods. Figure 99 shows growth enzymes E and E_p, transport RNA C, and pool P_n in these conditions. It appears that the growth is slightly unbalanced, but the system continues to grow. Since these experiments were also of interest from the point of view of cellular dormancy, where there is no growth or extremely slow cellular turnover, extended observation time was of particular importance. It is expected that once the cellular system has itself stabilized in the process of dormancy, the concentration of functional entities will be kept constant. Another important consideration here is that in order to demonstrate dormancy, it is essential to show that the cell is capable of maintaining itself at a low external pool value, which means that the flow of nutrients into the cell is very low.

Principal experiments are presented in Figs. 100–102. In these experiments, at about three-fourths of the normal "generation time," rate constant k_{17} was increased from the

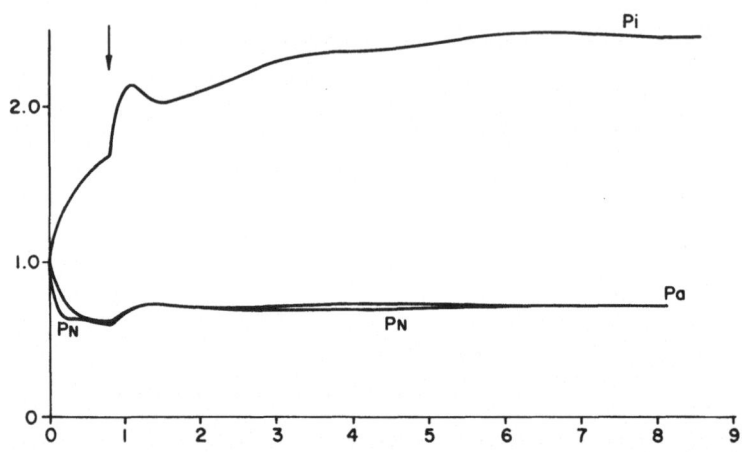

Fig. 101. Same experiment as in Fig. 100. Pools P_i, P_a, and P_n recorded.

value 1 to a value of 5.6. This k_{17} value was determined during the experiments on the computer, where k_{17} was varied as being a value which permits the system to grow very slowly. The following events take place when k_{17} is increased. Pools P_n and P_a both have a small transient increase, and, subsequently, a constant concentration is maintained. Pools are the first elements of the system to establish a steady concentration. Subsequently, transport RNA C, followed by messenger M and enzyme E, also establishes a rather steady value. After a

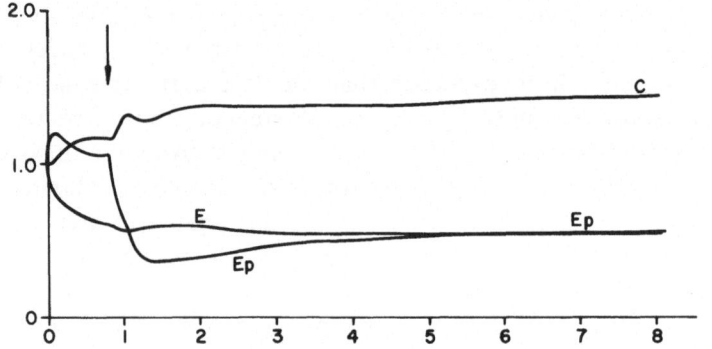

Fig. 102. Same experiment as in Fig. 100. Enzymes E and E_p, transport RNA C recorded.

more extended period, ribosome B and enzyme E reach steady values. A considerable while later, polymerase E_p and then pool P_i pass the transient phase. Templates N and N_p (not shown on graph) are the final elements to come to a steady value. It is evident that all functional entities finally establish themselves on a steady, but extremely slow growth. These experiments reveal that in the model-system, by variation of the k_{17} value, it is possible to control the long-term growth rate, and it suggests that in cellular growth polymerase inhibition provides potentially a method to exercise growth control. This mechanism could be of great importance in growth regulation, where various tissues have different replacement times. The principal characteristic of this type of control is that after the initial transient, the principal functional entities are brought effectively under control without any dramatic distortions which we have seen occur when control is estab-

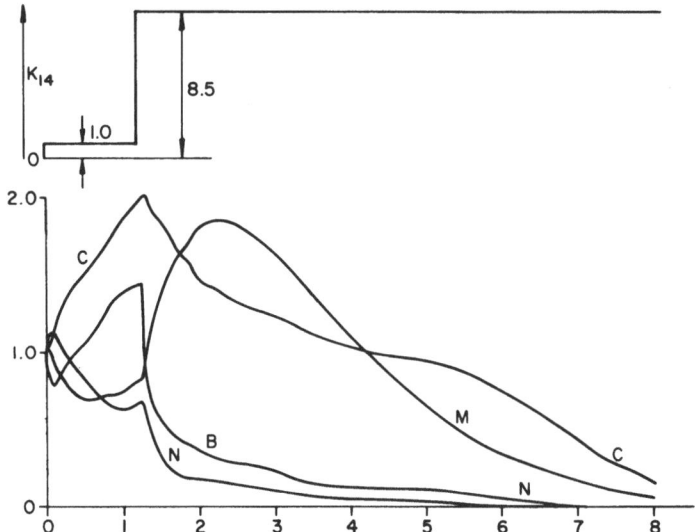

Fig. 103. The effect of ribosome B inactivation on growth. At time indicated on the graph, k_{14} is increased from value of 1.0 to value 8.5 and maintained at that level. Under these conditions system will become nonfunctional. Messenger M, ribosome B, transport RNA C, and template N are recorded.

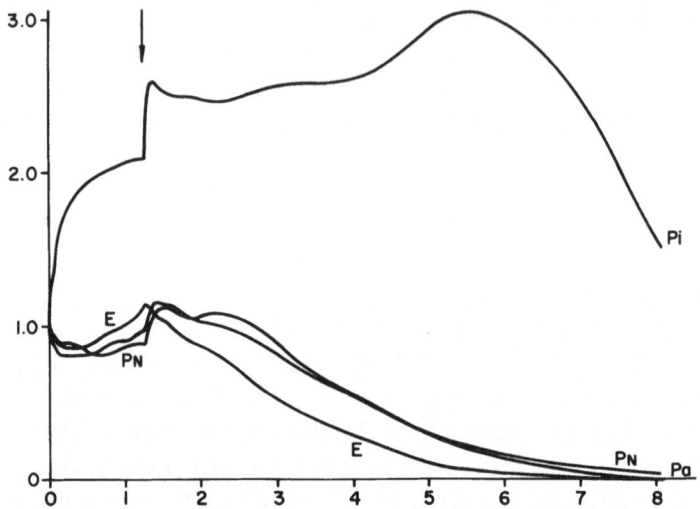

Fig. 104. Same experiment as in Fig. 103. Pools P_i, P_n, and P_a and enzyme E are recorded.

lished in various other sites in the system. While the template system here becomes initially large for a while, it gradually approaches a steady value.

Ribosomal Inactivation.   Ribosomes represent a major RNA fraction in the cell.   The specific inactivation of a ribosome is expected to produce very drastic results. Ribosome inactivation in the model-system can be carried out by varying k_{14} (equation 14, Table II).   The experiments shown in Figs. 103 and 104 show the results when a drastic inactivation is produced at the ribosome level. Here k_{14} is raised from a value of 1 to a value of 8.5. As expected, the first result of the increase of k_{14} is a rapid reduction of ribosome B (Fig. 103). This is followed by a reduction of templates N and N_p (not shown on graph).   Later it is to be expected because ribosome constitutes a part of the template. There is also initially a rapid reduction of transport RNA C, which is mostly caused by a regulatory step (equations 21 and 23, Table II). In equation 23, we shall see that part of E_p will be liberated from the $[E_pB]$ complex, since ribosome concentration becomes low. Consequently, there is a buildup of free E_p. This has been observed on the computer (not recorded). The increase of E_p will result in additional $[E_pC]$ complex formation. Therefore, there is a reduction of RNA C. In contrast, there is a rapid increase of messenger M, which is caused mainly by the availability of excess E_p, which rapidly takes up pool P_n (equation 6, Table II). The buildup of messenger M is temporary: it reaches a certain maximum value and subsequently starts to decline. Figure 104 shows that, after the increase of k_{14}, there is a rapid increase of pool P_i, since B decomposes via k_{14} into pool P_i. P_i concentration stays rather constant for a while, then starts to increase, reaches a maximum, and, finally, declines steadily. Pools P_a and P_n also have some initial transients, but there is subsequently a steady decline of both pools. Enzyme E will start to decline immediately after k_{14} increases, and finally reaches an extremely low level. Computer observations reveal that after the initial transient other components of the system also start to decline and approach the zero concentration level. The system becomes extinct because it is not able to compensate for such a high loss of ribosomes.

The next experiment was carried out to study the recovery from injury resulting from ribosomal inactivation. In this experiment, slight modification had to be made in the parameters of the system, since the inactivation of ribosome causes transients in the levels of some functional entities which are very large, especially in P_i and $[CP_a]$ complex. In order to keep these components within the operational limits of the computer, rate constant k_9 was reduced. This produces slight alterations in the system, and these experiments are not exactly comparable with the previous ones (Figs. 103 and 104). However, they are comparable in the sense that injury at ribosome level was exactly the same as in the previous case. It appears that there are some minor changes in the initial kinetics of growth. Experiment shows (Figs. 105 and 106) that, except for some minor changes, the effect produced by the inactivation of ribosomes is generally the same as in the previous experi-

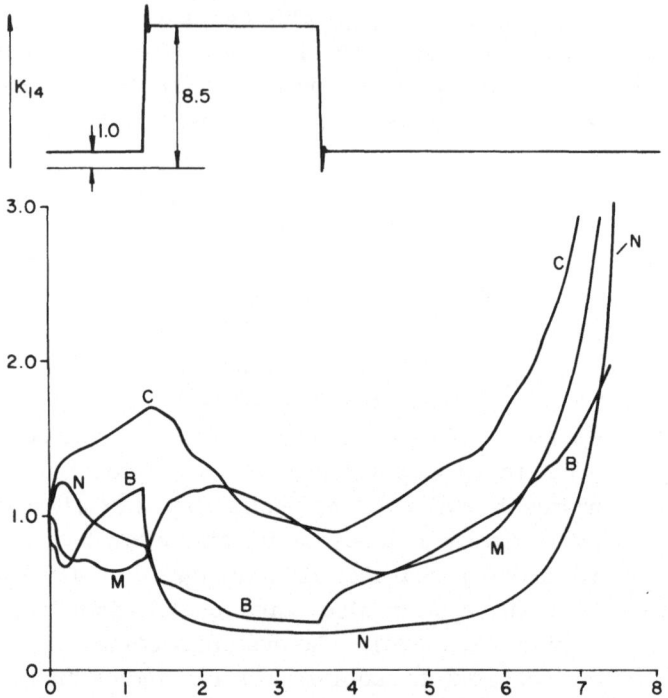

Fig. 105. The effect of ribosome B inactivation on growth. Ribosome inactivation occurs only for a certain period and is then terminated. During that period, k_{14} value is increased from value of 1.0 to 8.5. Under these conditions, system will recover from injury. Messenger M, template N, ribosome B, and transport RNA C are recorded.

ment. Therefore, we shall follow events when ribosomal inactivation is terminated by making k_{14} again equal to one. The first effect is an immediate increase in ribosome B concentration, followed by an increase in transport RNA C. Both B and C start to grow at practically the same rate. The next entities to increase are template N and enzyme E, followed by messenger M and pool P_n. After a while, there is an increase in P_a concentration, while pool P_i continues to decrease. Computer observations reveal that it takes few "generation times" before P_i starts to increase. A general kinetic character of cellular recovery after the termination of ribosomal inactivation is the initial, uneven, slow growth, followed by the adjustment period, after which normal growth finally takes place. The last condition is achieved after about seven "generation times." Further experiments were carried out on the computer to determine whether the increase of the length of time for ribosome inactivation would affect the (cellular) system's recovery. Experiments revealed that when the time of inactivation was extended beyond a certain limit, the system was not able to recover.

Messenger M Inactivation. Messenger M inactivation can be carried out on rate constant k_{12} level (equation 12, Table II). Experiments on the computer reveal that inactivation of M produces violent changes in the concentrations of many functional entities, especially in messenger M_p, enzyme E_p, and

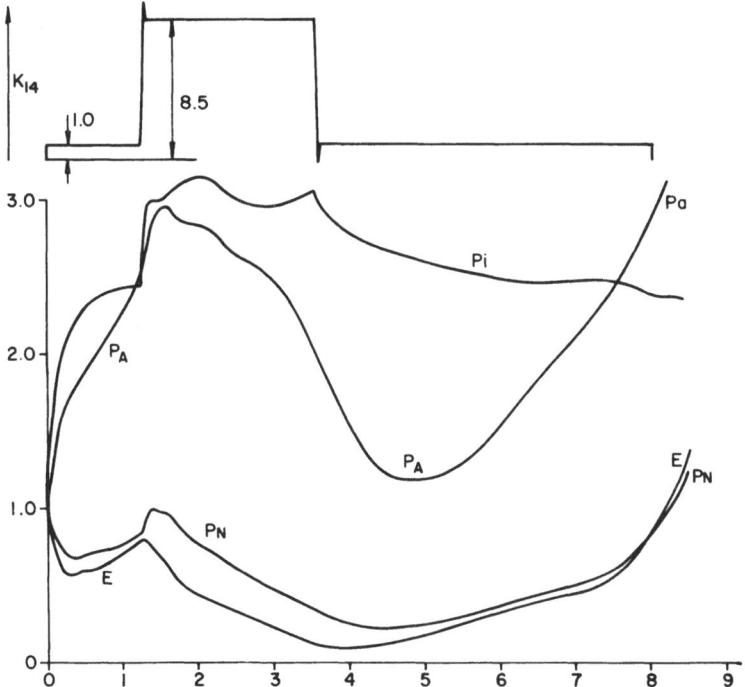

Fig. 106. Same experiment as in Fig. 105. Pools P_i, P_a, P_n, and enzyme E are recorded.

ribosome B. It is suggested that this type of cellular injury or loss of (cellular) viability would be associated with drastic changes in morphology. Furthermore, it was observed that inactivation of M is not a very effective way of destroying the functional system, because it can be rapidly resynthesized, and the system will recover. Thus effective activation on the M level is possible only when inactivation is very drastic and the duration of inactivation is long. Since in the injury and recovery processes concentration values of many entities are rather drastic, no recordings were made for these experiments.

Inactivation of Transport RNA C. Inactivation of C can be carried out on the k_{15} level (equation 15, Table II). When k_{15} is suddenly increased, there is a gradual decay of most of the functional elements, but there is a very rapid increase of pool P_a. This results from the lack of transport RNA C; consequently, the $[CP_a]$ complex cannot be built up, and free pool P_a becomes very large. This again produces some problems in computer technology, requiring rescaling of the program. Since these experiments do not yield any new type of basic information, only a limited number of observations were made directly on the computer. The conclusion is that the functional system can be effectively disorganized by inactivation of C.

Inactivation and Reactivation of Template N. In our previous studies, it was shown that the template is a suitable site for exercising growth control. One can pose the question: What would happen to the functional system if regulatory compounds such as hormones or substrates were applied on the templates in excessive amounts for long periods of time? Furthermore, are the effects on the system only temporary and always reversible, or would there possibly be permanent damage produced by the regulatory mechanisms when they are operational at an extreme level? Experiments for this purpose were carried out at first on the template N level. Template inactivation results when substrate s_i (equation 26, Table II) interacts with N and substrate s'_i restores template activity (Table II, equation 27). In the normal system, where there is unlimited growth for the model-system, rate constants k_{26} and k_{27} have zero value. Only when it is desired to control growth at the template level are rate constants k_{26} and k_{27} introduced. This mechanism is equally applicable to an external agent(s) introduced into the system, which causes specific template inactivation or activation. One can thereby demonstrate how an external agent can influence growth. In order to get the rapid inactivation and activation of the template, values assigned to rate constants k_{26} and k_{27} were relatively large. The introduction and the removal of substrates s_i and s'_i were synchronized again with the computer solution. Figures 107–110 show the effect of template inactivation and reactivation. Substrates s_i and s'_i are not present simultaneously, but operate separately. Obviously, one could carry out a number of experiments by varying many parameters of the system, such as the time interval when s_i is removed and s'_i is intro-

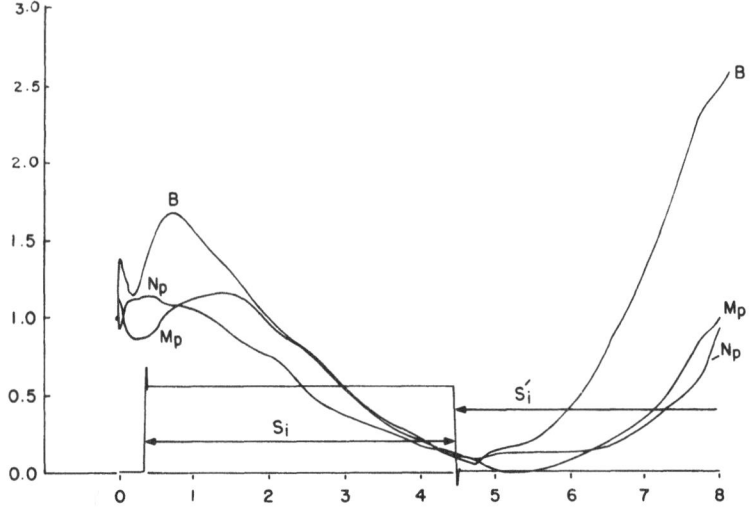

Fig. 107. The effect of activation and deactivation of template N. The template N is exposed to s_i, the inactivator substrate for a certain length of time. Subsequently the s_i removed from the system and s'_i the activator substrate, is introduced. System will recover from the injury. Ribosome B, template N_p, and messenger M_p are recorded.

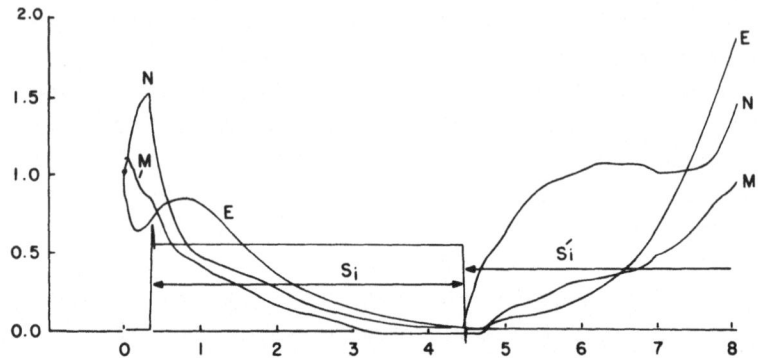

Fig. 108. Same experiment as in Fig. 107. Template N, messenger M, and enzyme
E are recorded.

duced in the system. However, since computer solution time
is limited, it is preferable that there be no time interval between the introduction and removal of those two compounds.
As seen in Fig. 108, at the time when s_i is introduced into the
system there is immediately a rapid reduction of active template N, which is converted into inactive form (N'). Messenger M, which is still in the declining phase of normal growth

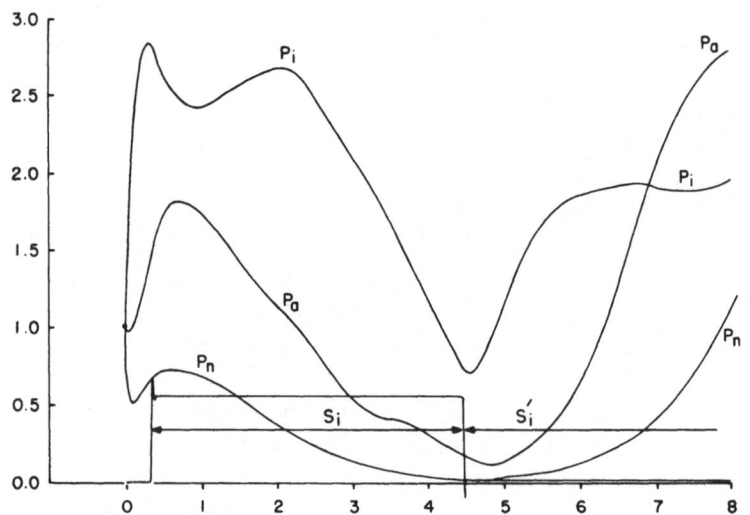

Fig. 109. Same experiment as in Fig. 107. Pools P_i, P_a, and P_n are recorded.

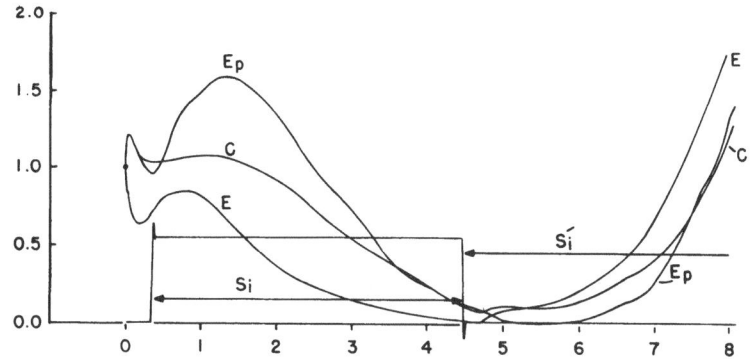

Fig. 110. Same experiment as in Fig. 107. Enzymes E_p and E and transport RNA C are recorded.

when s_i is introduced, starts to decline even more rapidly. All other entities of the system continue to grow for a short while, reach a certain maximum value, and then start to decline. The time sequence of decline for the entities is as follows: enzyme E, ribsome B, pools P_a and P_n, template N_p, polymerase E_p, transport RNA C, messenger M_p, and finally pool P_i.

When s_i is maintained in the system, all functional entities approach zero and the system decays completely. However, if at a certain time s_i is removed and s'_i is introduced into the system, computer observations reveal that the system may or may not recover, depending on the length of time that s_i is maintained in the system. This phenomenon is of particular interest from the point of view of the viability concept, and further studies on the subject will be carried out subsequently. In this experiment, s_i application time was selected to be short enough to permit recovery of the system. Let us observe the sequence of events, when s'_i is introduced into the system and when s_i is removed. The first step is the rapid buildup of template N. This is followed by the increase of pool P_i, ribosome B, messenger M, and enzyme E. After some delay, pools P_a and P_n start to increase, followed by the polymerase synthesis system, messenger M_p, template N_p, and enzyme E_p. Template N concentration reaches a maximum, declines slightly and after a short while starts to increase

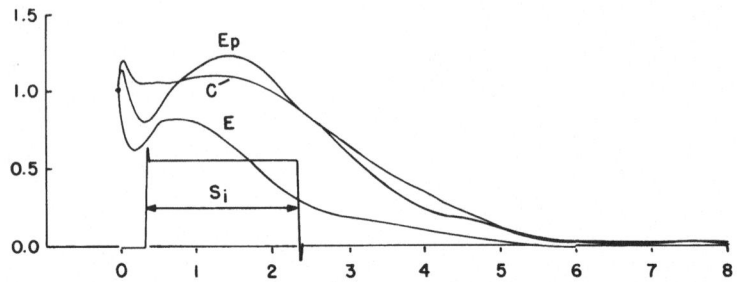

Fig. 111. The effect of template N inactivation. The substrate s_i is introduced in the system and is removed after a certain time interval. System is not viable under these conditions. Enzymes E and E_p and transport RNA C are recorded.

again. It is evident that the initial phase of reorganization is very uneven, but finally all components of the system will start to grow normally.

It was of further interest to carry out an experiment where s_i is removed from the system as before, but in which activating substrate s'_i is not introduced into the system. This means that the system, after being exposed to s_i for certain time, is left to its own fate. This experiment is shown in

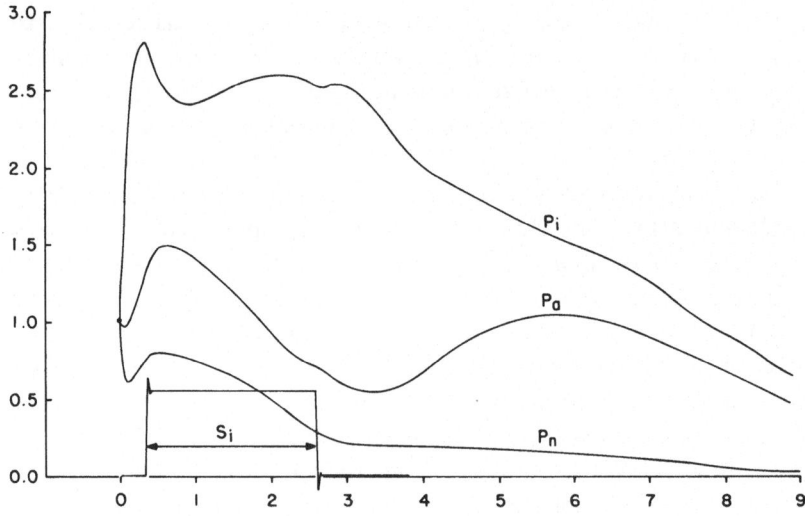

Fig. 112. Same experiment as in Fig. 111. Pools P_i, P_a, and P_n are recorded.

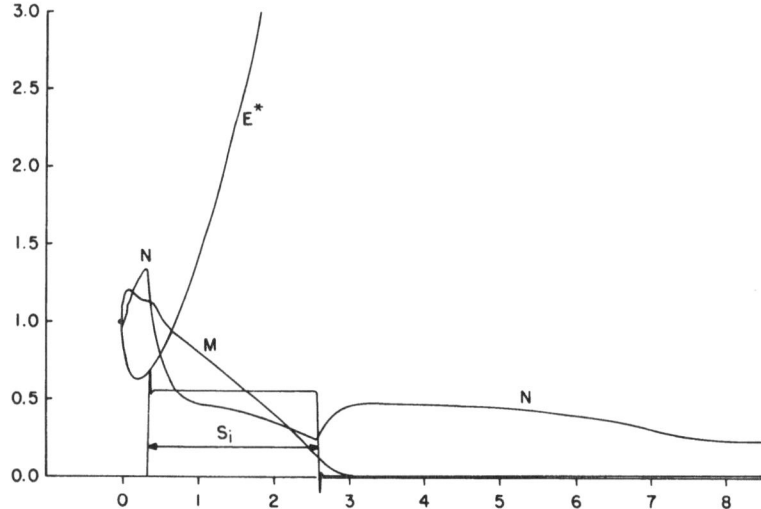

Fig. 113. Same experiment as in Fig. 111. Template N and messenger M are recorded. E* indicates normal growth without inactivation of template.

Figs. 111 to 114. Since long time periods are required for the growth transients to be resolved, s_i in those experiments is introduced by activating rate constant k_{26} in a relatively early phase of growth. For the purpose of orientation, normal growth of enzyme E is represented in Fig. 113 and is indicated by E^*. It appears that normal growth is rapid compared with the growth under restricted conditions. Substrate s_i is removed from the system at the same time as in previous experiments (Figs. 107—

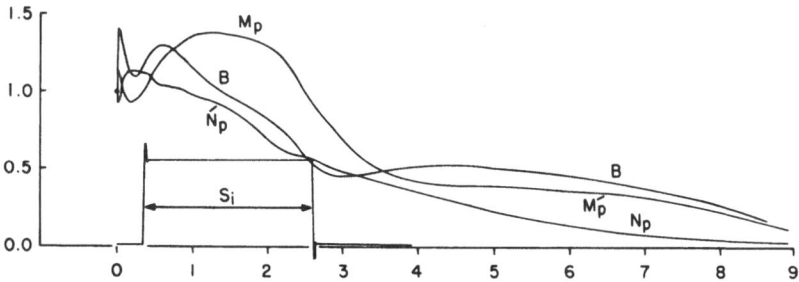

Fig. 114. Same experiment as in Fig. 111. Template N_p, messenger M_p, and ribosome B recorded.

110).   Since the effects of s_i are similar to those described previously, they are not described here. When inactivation of the template N is terminated by making k_{26} equivalent to zero again, there is an immediate increase of N concentration. N continues to be relatively constant for a while, and then starts to decline steadily.   Messenger M concentration continues to decline after the removal of s_i and finally reaches a very low level, while messenger M_p (Fig. 114), which also continues to fall, reaches a rather constant level for a while but finally slowly declines again.   Template N_p continues to decline in spite of removal of s_i.   On the other hand, ribosome B concentration continues to pass through a minimum, rises to a maximum, and subsequently starts to decline again. Figure 111 shows that enzyme E_p, after removal of s_i, continues to decline, but the rate of decline is reduced until a rather constant concentration level is reached, which is maintained for a long period, and then E_p starts to decline again. Transport RNA C continues to decline steadily and gradually reaches a very low concentration level. Enzyme E also continues to decline for a while, passes through a minimum, and then, after a slight increase, finally declines again. The behavior of the pools is of particular interest. At the time of removal of s_i, the concentration of pool P_i is still at a relatively constant level. However, shortly afterward P_i starts to decline steadily and continues to do so until the end of the experiment. In contrast, P_a also continues to decline after removal of s_i, but after passing through a minimum it starts to build itself up; it reaches a maximum, and after that there is a steady decline. Pool P_n continues to decline, is then rather constant for a while, and then declines very slowly.

It is of interest to note that the system makes an attempt to recover. Several functional entities have a secondary maximum (P_a, B, N, and E), but fail to sustain it. As a consequence, the system becomes extinct. It is not clear which of the functional entities cause the disorganization.   It appears that messenger M is the first to reach an extremely low concentration, and perhaps the M is the most limiting factor for the recovery of the system. However, it is surprising that at the time when s_i is removed, all functional elements are still present in the system. Nevertheless, the system does not recover its functional characteristics. Perhaps it can be interpreted in terms of an imbalance or disorganization developed in the concentration of various functional entities. It is instructive to observe the difference between this experiment and the previous one (Figs. 107–110), in which the system recovered in spite of the fact that the inactivation time was much longer. It appears that the introduction of the template activator s'_i indeed produced significant results in recovery. One of the principal symptoms of nonrecovery seems to be the inability of the system to build up pool P_i. Obviously, the system will be terminated if P_i continues to decline steadily as a function of time. It further appears that the conversion of inactive templates into active templates by s'_i is essential. The synthesis of enzyme E could thus occur and external pool P_e could be converted into internal pool P_i. The very fact that the system cannot establish an effective transport system, able to transfer external pool to internal pool, could indeed be a limiting factor in this experiment. Thus, we can conclude that, since all entities are present, the reason for functional disorganization of the system is not the lack of functional entities, but rather the unbalanced distribution of entities. The principal effect of substrate s'_i is its ability to convert inactive templates into active templates, thus enabling a rapid protein synthesis, and thus reversing the trend of the decay of the system.

These model-system experiments suggest that an excessive application of a growth regulator for too long a time can extinguish a cellular system. It appears that when extensive growth retardation has taken place or a growth decline has actually occurred, only the application of the growth promoter will save the cell. Obviously, in the cellular system the regulatory mechanisms are more complex than those in the model-system. However, the underlying principle involved is the same for both cases. It is only the degree of complexity

of organization which is different in the two systems. The
principle seems to be well established that any system can
disorganize if this process is continued long enough and if it
passes a certain critical point of self-recovery. In this framework, the question will arise as to whether it is possible even
for the application of a second activating substrate to be effective when the cell is disorganized extensively. In order to carry
out this experiment, a system which was growing at a very
slow rate was used. This type of system is more sensitive in
reflecting alterations in the functional entities much sooner
than the normal system.   This type of experiment is represented in Figs. 115 and 118. Functional entities indicated by
a star are the components during normal growth. Here, all the
components have reached a slow, steady growth, except E_p
polymerase, which has just reached its lowest plateau at the
end of recording time, and observations reveal that it will start
to grow from this point on. Furthermore, in order to speed up

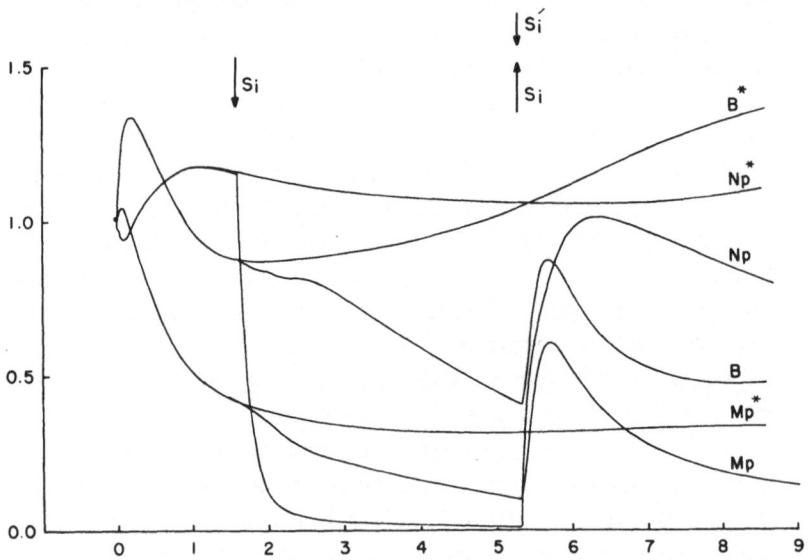

Fig. 115. The effect of template N and N_p, inactivation and subsequent reactivation.
System does not recover. Entities designated by stars are components of normal
growth. Template N_p, messenger M_p, and ribosome B recorded.

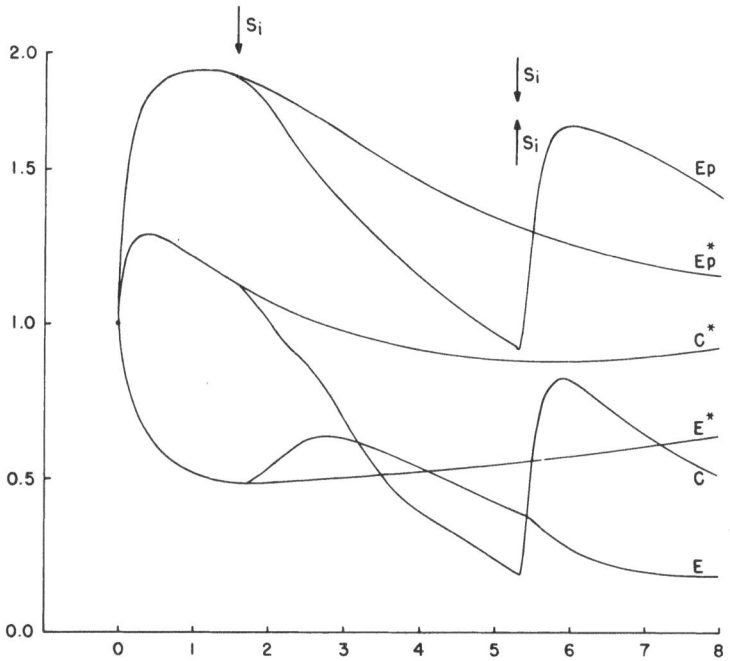

Fig. 116. Same experiment as in Fig. 115. Enzymes E and E_p and transport RNA C recorded.

the processes, both templates N and N_p were converted simultaneously into inactive and active form. When inactivator s_i is applied, there is immediately a very rapid reduction of templates N and N_p, followed by a slow decline of ribosome B, enzyme E_p, messenger M_p, and transport RNA C. While normal pools P_a and P_n are growing very slowly, P_n starts to decline shortly, while P_a passes through a few compensatory adjustments and continues to maintain its average concentration. In contrast, there is a slight increase of enzyme E and messenger M concentration. Both reach a maximum and then start to decline P_i is not recorded here, but it has similar characteristics, as shown in Fig. 112. When s_i is introduced, P_i is also reduced, and, as a result, there is a release of E from the complex [EP_i] (equation 22, Table II). This compensatory adjustment causes an initial increase in enzyme E concentration.

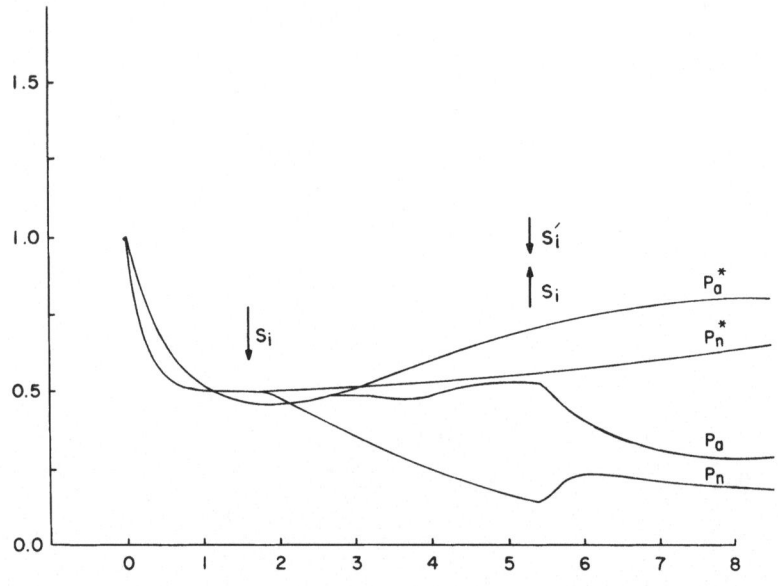

Fig. 117. Same experiment as in Fig. 115. Pools P_a and P_n recorded.

If this experiment is continued while s_i is maintained in the system, all entities will decay finally to a very low level and the system will cease to exist. Therefore, an attempt was made to reactivate the system by introducing the activating compound s'_i for the purpose of converting inactive templates N' and N'_p into active form (equations 25 and 27, Table II). After the introduction of s'_i and the removal of s_i, there is a rapid increase of N, N_p, E_p, C, B, and M_p. However, all these elements, after reaching a maximum, sooner or later start to decline. Computer observations reveal that template N and pool P_a also reach a maximum at a considerably later time and then start to decline, while pool P_n, after an initial increase, continues to decline steadily. Enzyme E and messenger M are continually declining. After the initial increase of many entities, there seems to be an apparent recovery of this system. However, some entities seem unable to organize themselves, and will decline continuously until concentrations approach the zero value. Principal entities which decrease persistently after application of s'_i are enzyme E and messenger M. Those two entities probably also cause a decline of other entities, and the system becomes nonfunctional. This seems to be a borderline case because experiments on the computer reveal that when inactivation time is slightly reduced, this system becomes viable. One can project, on the basis of these experiments, that the phenomenon of cellular recovery is highly conditional. It can be affected by a multitude of factors, such as variations in the stability of functional entities as well as the rate of their synthesis.

In order to study the self-recovery of the system, the degree of injury was reduced and the number of sites being inactivated was limited to one. It is of interest to explore what factors and conditions affect the self-recovery of the system. In these experiments, only template N was inactivated. Figures 119 and 120 show the results. It is indicated that when s_i is introduced into the system, there is a rapid reduction of template N and messenger M. Enzyme E increases, passes through

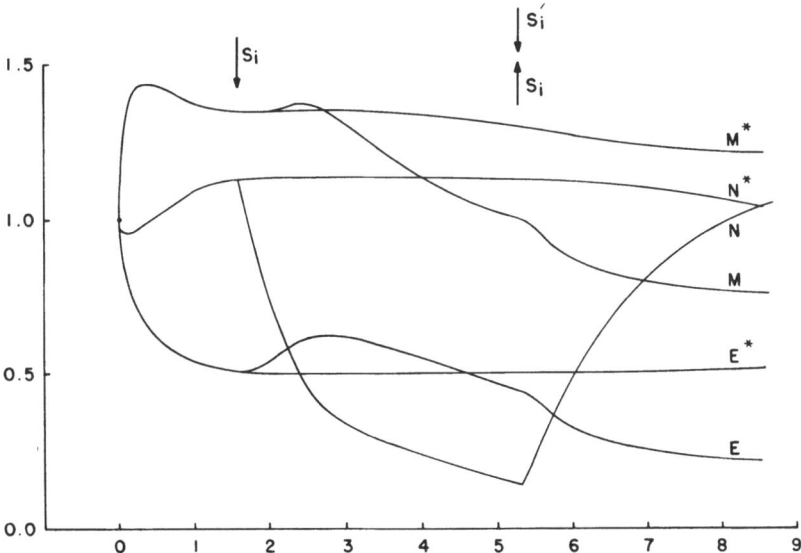

Fig. 118. Same experiment as in Fig. 115. Messenger M, template N, and enzyme E recorded.

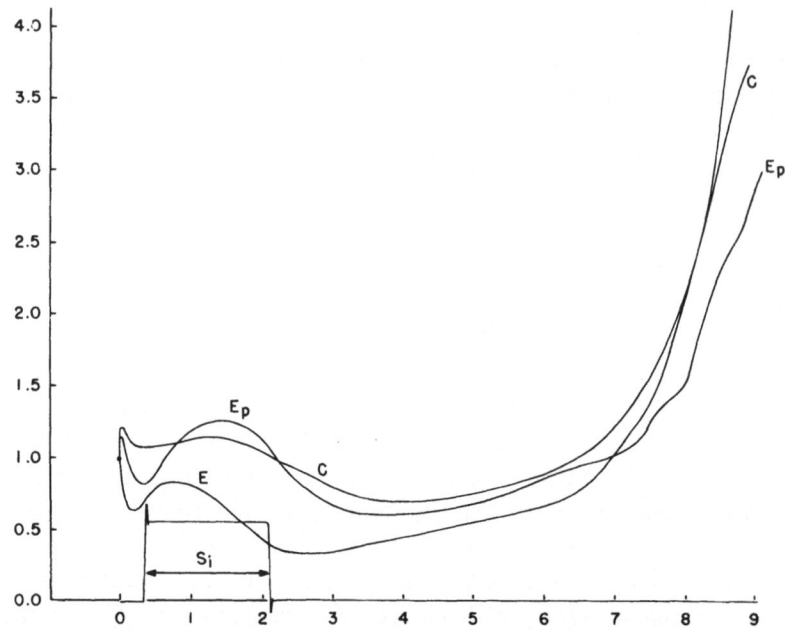

Fig. 119. The effect of temporary inactivation of template N. System will recover
from injury. Enzymes E and E_p and transport RNA are recorded.

a maximum, and then continues to decline. Enzyme E_p, transport RNA C, and ribosome B possess generally similar characteristics. At the time when s_i is removed from the system, all components are declining. Computer experiments reveal that if s_i is kept in the system, it will become nonfunctional. However, the removal of s_i from the system produces various changes in the system. The first event is the increase of template N concentration, followed by messenger M, which increases very slowly but steadily. Enzyme E continues to decrease, passes through a minimum, and starts to increase slowly. It appears that the entities required for enzyme synthesis are the first to become operational. Ribosome B also starts to increase relatively soon, while transport RNA C and enzyme E_p exhibit considerable lags in growth. It is evident that there is an adjustment time which lasts several generation times before the system will grow normally again. It was of interest to determine the length of time during which the system can sustain the damage. Therefore, the experiment was carried out to determine how long s_i had to be maintained in the system so that it would not recover again. Computer observation revealed that when inactivation time was increased by about 13%, the system was unable to recover. Figures 121 and 122 reveal the borderline viability experiment. It is evident that template N has a transient increase after the removal of s_i, stays constant for a while, and then starts to decline. Messenger M remains practically constant for a while and then starts to decline. Ribosome B exhibits a slight change in decay slope, but continues to decline. All other entities continue to decline. It is of particular interest that elements required for enzyme E synthesis were unable to establish an increasing trend of growth. This experiment again reveals the importance of the enzyme forming system in recovery. It demonstrates that it is sufficient to increase the inactivation time by a small increment in order to disorganize the system thoroughly.

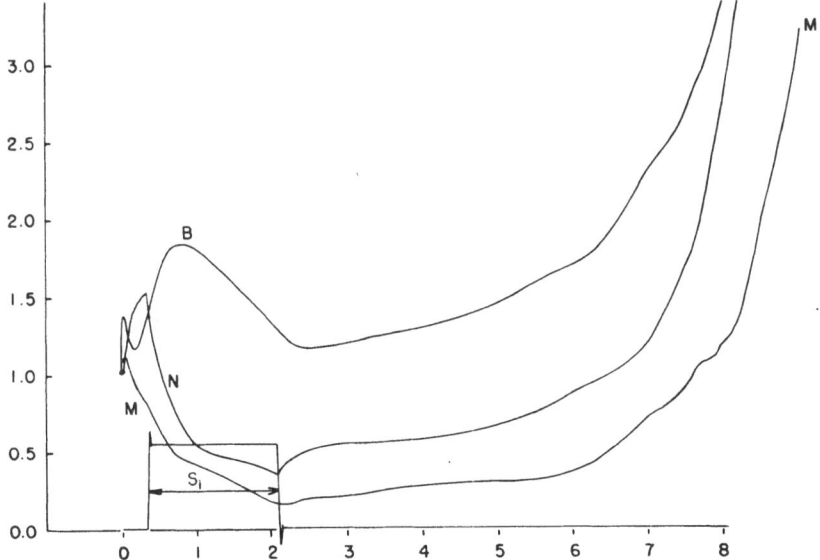

Fig. 120. This is the same experiment as in Fig. 119. Messenger M, template N, and ribosome B are recorded.

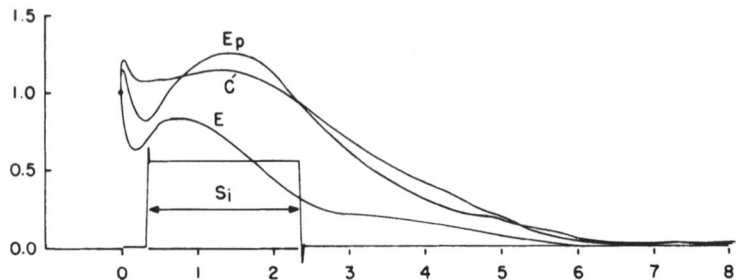

Fig. 121. The effect of temporary inactivation of template N. This is the same experiment as in Fig. 119, except that the application time of s_i is longer. In this case the system does not recover. Enzymes E and E_p and transport RNA C are recorded.

It is of interest to know whether the ability of the cell to re-cover depends on the stability of certain functional entities. Since, in the experiments shown in Figs. 121 and 122, the protein-forming system involving messenger M, template N did not show positive growth, the question is raised whether the system's recovery could be helped, when the stability of messenger M would be increased. This experiment was carried out and explorations on the computer revealed that when messenger M decay constant k_{12} was sufficiently decreased, the system recovered. Figure 123 shows some of the entities in

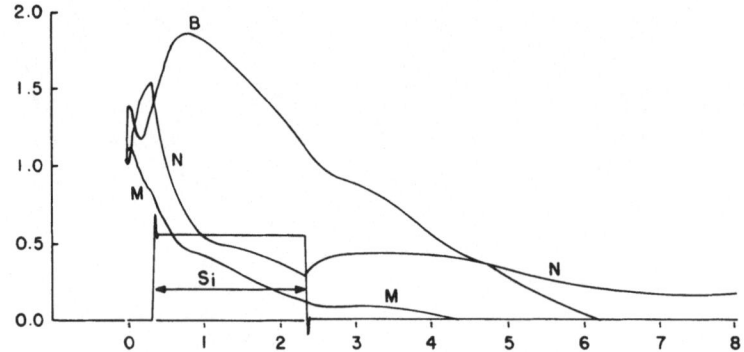

Fig. 122. This is the same experiment as in Fig. 121. Messenger M, template N, and ribosome B are recorded.

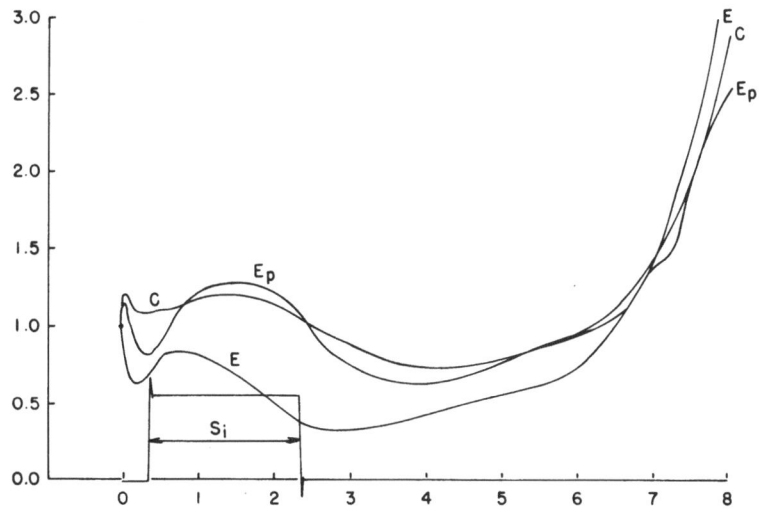

Fig. 123. The effect of temporary inactivation of template N. This is the same experiment as in Fig. 121, except that messenger M stability increased (k_{12} value is decreased from 0.0029 to 0.0010). In this condition, the system is again viable. Enzymes E, E_p, and transport RNA C are recorded.

this experiment.   Here, rate constant k_{12} was reduced 66%.   It is evident that enzyme E and E_p and transport RNA C, after removal of s_i, start gradually to increase. Computer observations reveal that all other functional entities will increase finally and that the system will grow normally.   The evidence that the stability of an entity can affect the viability suggested further studies in this area. However, this represents an extensive research problem, since many factors enter into the system's recovery process and a systematic study would, of course, be required.   Here we shall report only a few of the principal findings. An interesting observation was made when it was decided to study the effect of the synthesis of ribosomal RNA fraction (B') by increasing rate constant k_2.   Otherwise, the experiment was carried out in exactly the same conditions as previously.   This experiment revealed that the increase of k_2, instead of leading to a more rapid recovery of the system, produced reverse effects.   Experimental results are shown in Fig. 124.   It is evident that the system, instead of making a

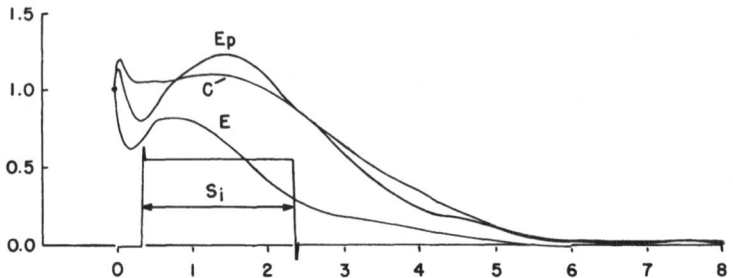

Fig. 124. The effect of temporary inactivation of template N. This is the same experiment as in Fig. 123, except that the rate of ribosome precursor B' formation is increased. (k_2 value is increased from previous value 0.504 to 0.564.) System is not viable in this condition. Enzymes E, E_p, and C are recorded.

more rapid recovery, is becoming nonfunctional. This occurred in conditions where the increase of rate constant k_2 was 12%. This result seems surprising, since one could expect that the increase of the rate of synthesis of a functional entity should improve the ability of the system to recover. However, the experiment reveals the opposite effect, since all functional entities finally decline. Here, more systematic studies would be required to explore this problem in detail. However, it appears that the system decays because of excessive ribosomal RNA formation which depletes the pool P_n, which is required for messenger M formation. This shows that under conditions in which there is a limited supply of pools available, the competitive reactions may affect the survival of the system. It was then asked whether the system which is represented in Fig. 124 again could be made viable while a high rate of ribosomal synthesis was continued. Exploratory experiments were carried out which indicated that the easiest way to help this type of system to recover is to increase the protein stability. A viable system was produced when enzyme E decay constant k_{16} was decreased about 12%. Figure 125 reveals that enzymes E and E_p and transport RNA C are, after a certain delay, all increasing again, and normal growth follows. Observation reveals that all entities of the system are recovering.

In summary, experiments reveal that the ability of the system to recover is highly conditional. It depends on the interaction of various functional entities, and, while certain relations produce a viable system, other relations may be detrimental to its recovery. Small parametric changes determine the system's ability to recover. This fact suggests that in cellular systems under borderline conditions recovery may be extremely slow and may depend on any minor variation in the environment. On the computer, it depends on the noise level within the electronic circuitry.

Injury to Cell Membranes. In order that cellular growth occur, it is essential that the integrity of the functional system be maintained.   This implies that the cell membrane which isolates the intracellular system from the environment has to

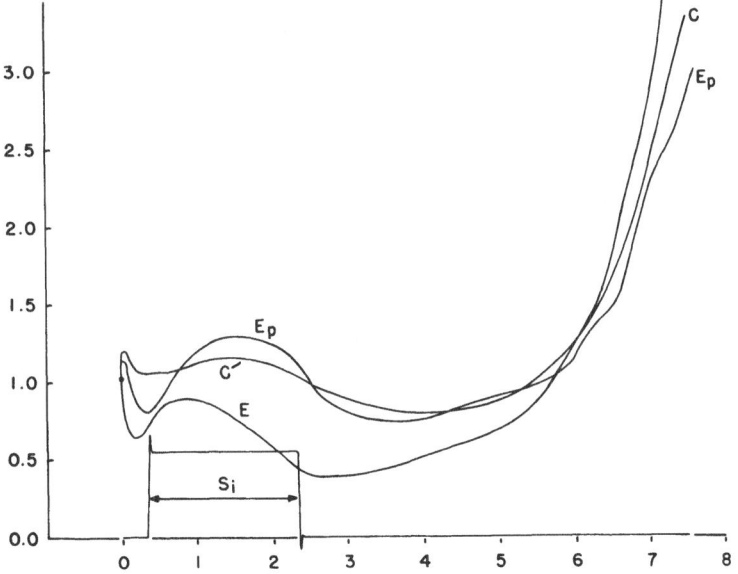

Fig. 125. The effect of temporary inactivation of template N. This is the same experiment as in Fig. 124, except that enzyme E has been made more stable. For this purpose rate constant k_{16} has value decreased from 0.068 to 0.060. The system has again become viable.   Enzymes E, E_p, and transport RNA C are recorded.

be intact and functional. It not only contains a transport system which not only brings external nutrient medium into the cell, but which also permits some components of intracellular metabolities to escape into the external medium. In the model-system, cellular leakage is simulated by the rate constant k_{30}, by which a small fraction of internal pool P_i escapes and is transformed into a compound X, thus indicating that there is no re-entry of pool into the system. Factors which increase leakage in the model-system by increasing k_{30} will be expected to modify the behavior of the functional system. It is well known in cellular biology that there are a variety of compounds which affect cell membranes, such as cationic detergents, polypeptides, polysaccharides, etc. The action of various compounds on the cell membrane has been

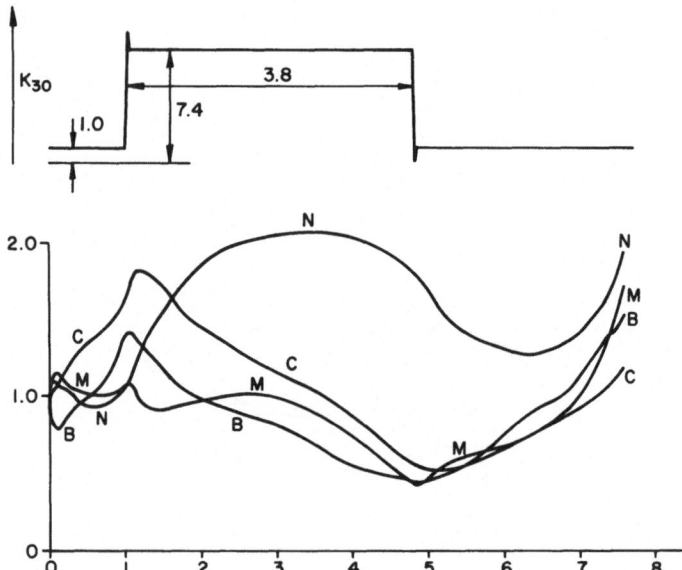

Fig. 126. The effect of "leakiness" on viability of the system. At a certain time k_{30} is increased from the value 1.0 to the value 3.8, and for a certain duration this condition is maintained. Subsequently k_{30} is reduced back to value 1.0. System will recover from this type of injury. Pools P_i, P_a, and P_n and enzyme E are recorded.

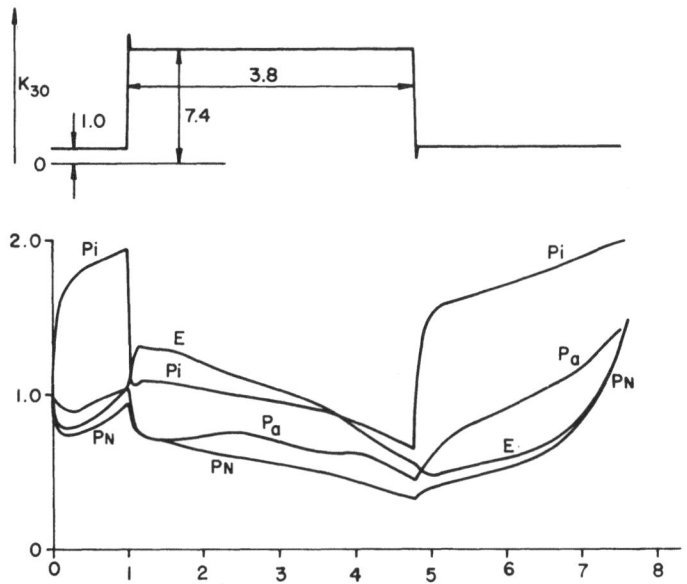

Fig. 127. The effect of a leaky membrane on cell recovery. This is the same experiment as in Fig. 126. Template N, messenger M, ribosome B, and transport RNA C are recorded.

reviewed by Davis and Feingold [11]. It is evident that some of the compounds produce extensive membrane damage, allowing intracellular contents to escape into the medium. If this leakage is extensive, complete lysis can take place and cell becomes nonfunctional.

In order to carry out simulation experiments on the computer, it was essential that a slowly growing cell be used, since the rate changes of various processes are very slow in this type of injury. Consequently, external pool P_e was made smaller and k_{30} was made larger, so that we had initially a leaky, slowly growing cell. In these experiments, observations were normally carried out during nine "generation times." The leaky membrane was simulated in computer experiments by changing k_{30}. At certain times it was increased above the normal value, and at a certain later time it was decreased back to the normal value. The questions were raised as to how long the system can sustain an excessively leaky membrane and still be viable, and whether there is a recovery from conditions where leakiness has lasted for a long time. The first experiment demonstrates the effect of temporary leaky conditions on the growth of the system (Figs. 126 and 127). An initially growing system had a relative k_{30} value of one. At a certain time, k_{30} was increased to 7.4, and after a certain time k_{30} was reduced back to its normal value. When k_{30} was increased, there was at first a rapid reduction of pool P_i, and at the same time there was an increase in the concentration of enzyme E. This is produced by the regulatory compensation in which part of E is liberated from $[E\,P_i]$ complex (equation 22, Table II). Rapid reduction of P_i also causes reduction of pools P_a and P_n. All pools continue to decline when k_{30} has a high value. Similarly, enzyme E is declining continuously. In contrast, messenger M, ribosome B, and transport RNA C will continue to increase, reach a maximum, and then start to decline. However, messenger M goes through a secondary maximum before it starts to permanently decline. A remarkably different result is indicated by template N, which continues to increase for a long time, reaches a maximum, and finally starts to decline. When k_{30} returns to normal value, there is a rapid rise of pools, followed by a steady growth. There is initially a regulatory reduction of enzyme E, but gradually its growth rate increases. There is an immediate rise of messenger M, followed by ribosome B and transport RNA C. Template N continues to decline, passes through a minimum, and then starts to grow rapidly.

The experiment reveals that the system can recover from a leaky condition which lasts several "generation times." The question was raised whether the increase of the duration of leakiness would produce a set of conditions under which the system is incapable of recovery. Exploratory tests on the computer revealed that when duration of leakiness was increased from 3.8 to 4.2, the system was incapable of recovery. This experiment is shown in Figs. 128 and 129. When k_{30} is made equal to unity, pool P_i again rises rapidly, maintains for a while a nearly constant level, but finally starts to decrease. Pool P_a also increases rapidly initially, the growth rate is then reduced, and the concentration finally reaches a maximum value. Subsequently, P_a starts to decline slowly. Pool P_n increases very slowly. Enzyme E, after initial reduction, stays practically constant for a while and then starts to reduce slowly. Messenger M and transport RNA C continue to decline, while ribosome B starts to increase. When observation time is expanded on the computer, it will be seen that all components finally approach zero. It is evident that a roughly 10% increase of duration of leaky conditions was sufficient to disorganize the system permanently. Data presented in the simulation experiments (Figs. 126–129) suggest that the exposure time represents a very important element in determining cellular viability.

Next, we shall consider that the system is exposed to the same leaky conditions as in Fig. 128 and 129, but that some

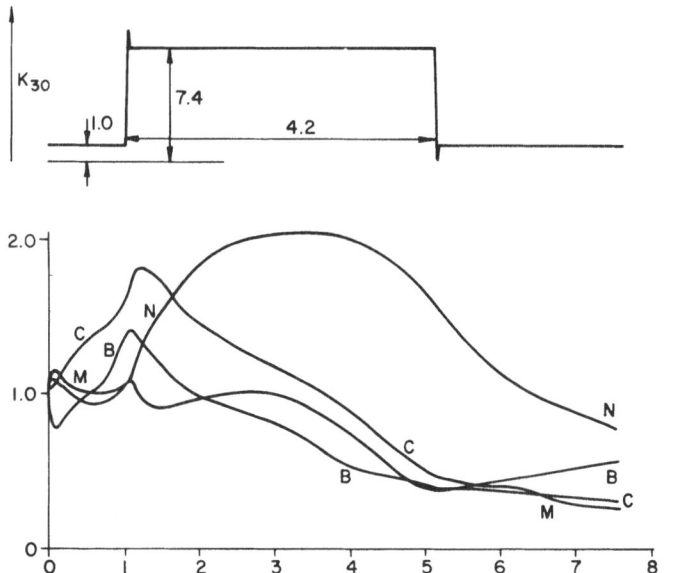

Fig. 128. This is the same experiment as in Fig. 126, except that the relative duration of "leakiness" has been increased from 3.8 to 4.2. In these conditions the system is not more viable. Template N, messenger M, ribosome B, and transport RNA C are recorded.

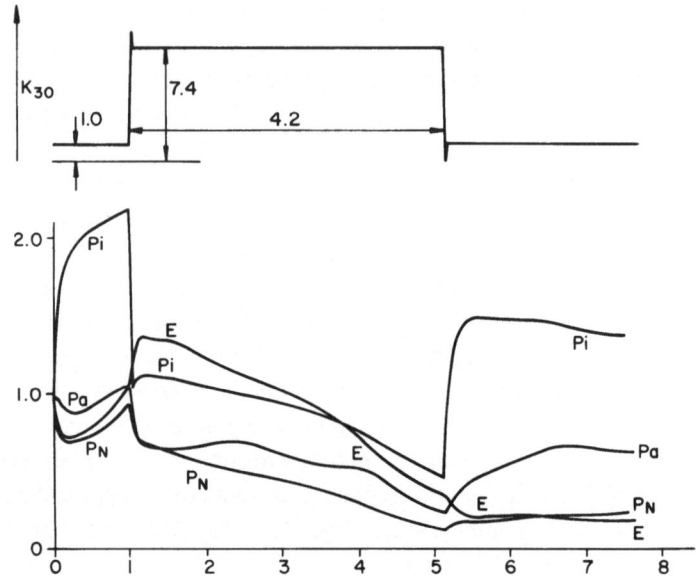

Fig. 129. The effect of a leaky membrane.   This is the same experiment as in
Fig. 128. Pools P_i, P_n, and P_a and enzyme E are recorded.

parametric value of a functional entity is modified until a
system results which will be able to maintain its functionality.
Exploratory experiments on the computer reveal that, in order
to produce a system which remains viable, it is sufficient
to reduce the k_{12} value which characterizes the ability of messenger M, roughly by 2%. Experiments are shown in Figs. 130
and 131 in which k_{12} was reduced by 2%, while other conditions
are exactly the same as in the previous experiment (Figs. 128
and 129).

It is evident that after the increase of k_{30}, all entities
behave similarly, as described previously (Figs. 128 and 129).
However, after termination of the leaky condition, there is an
increase of all functional entities. A similar growth sequence
was observed in previous experiments, where recovery of the
system took place. It is considered of great significance that
a general, nonspecific injury such as membrane leakiness,
which would not affect any particular functional entity, might be

capable of destroying the system, depending on a small variation in stability within a functional element. In order to explore
this aspect of the problem, further studies were performed.
The system was put into the condition of no recovery, as
indicated in the experiments shown in Figs. 128 and 129, and
the following rate constants were varied in relatively large
range: k_{15}, which characterizes transport RNA stability; k_{14},
which characterizes ribosome stability; k_{17}, which characterizes E_p polymerase stability; and k_{13}, which characterizes M_p
stability. None of the changes were effective in helping the
restoration of functionality. However, the rate constants k_{12} and
k_{16}, which represent the stabilities of messenger M and enzyme E, respectively, were effective for the recovery. Here
again, as in several previous experiments, it is indicated that

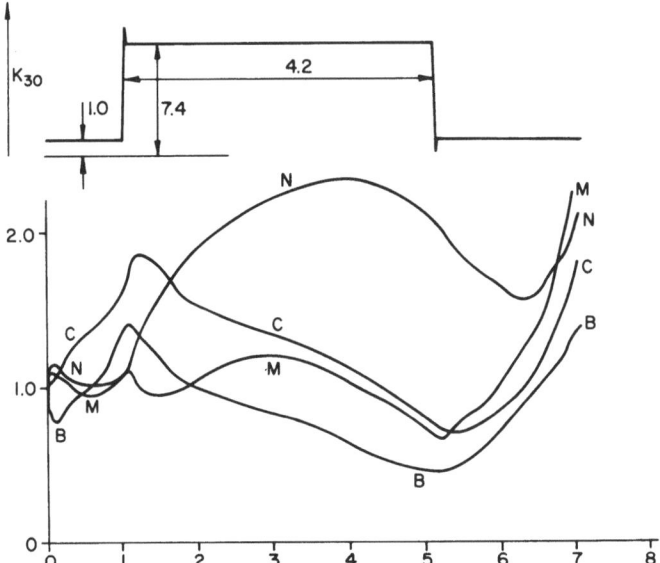

Fig. 130. The effect of leaky membrane on cell viability. The same experiment as
in Fig. 128, except that messenger M stability has been increased. Rate constant
k_{12} has been decreased from the value 0.9700 to 0.0950. As a consequence the system has become viable. Template N, messenger M, ribosome B, and transport
RNA C are recorded.

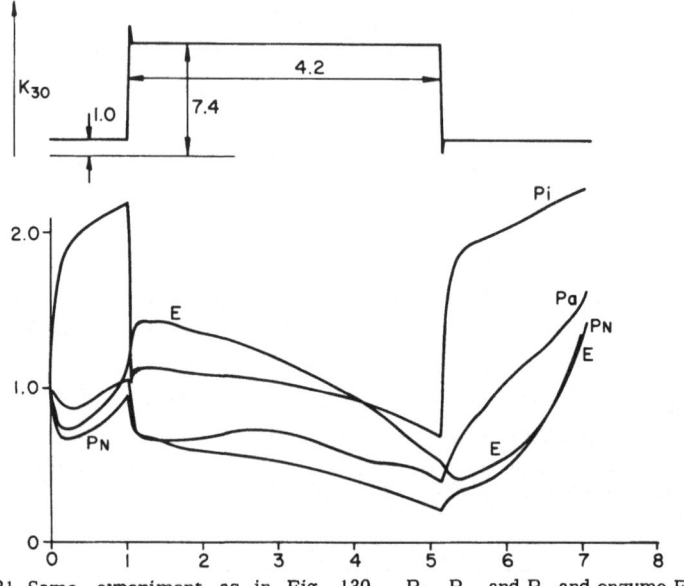

Fig. 131. Same experiment as in Fig. 130. P_i, P_a, and P_n and enzyme E are recorded.

the operation of the protein-synthesizing system is a most crucial one for the recovery process.

Inactivation of External Pool Transport System.   Cell growth depends on the availablity of an external nutrient source, and any factors which interfere with transport of this source will affect the growth rate.   For example, when either specific enzymes (which are operative in the transport mechanism) or nonspecific permeability pathways are affected by external or internal agents, the result is a reduction of the transport rate. The principal object here is to analyze the effect of impairment of the transport process as an injury mechanism. On the other hand, since various functional processes can be analyzed from different points of view, we shall now also further extend studies on the regulatory mechanism for cell growth. Simulation of cellular injury at the level of transport mechanisms, where external agents will affect the transport enzymes, can be studied on the model-system where external pool P_e is converted to pool P_i by enzyme E_t. By reducing the enzyme activity, the injury effect can be simulated. On the other hand, in order to study growth control, one has to assume that the transport system is under a regulatory control. Substances which trigger the control mechanism can operate from within the cell, or these can act as agents from outside. For example, hormones would be normal extracellular growth-regulating agents, while a surface-active chemical compound which interacts with transport enzyme could be an agent producing cellular injury. The experiments were so organized that two of the problems could be analyzed at the same time. Assume that an agent A, which can be a regulatory or an inactivating component, interacts with enzyme E. The fraction of enzyme E which is active as a transport enzyme is $k_t E$. Rate constant $k_{18} = k'_{18} k_t$ (Table II) represents transport enzyme activity in terms of total enzyme concentration. The specific transport property can be modified when k'_{18} is altered. Assume that $A + E \rightarrow E'$, where the activity of E' is characterized by k'_{18}, which determines whether the compound is an inhibitory or activating type. Transport of the external pool into the internal pool proceeds via:

$$P_e \xrightarrow{k''_{18} k_t} P_i$$

By variation of k''_{18}, pool transport can be affected. In order to simulate the cellular growth in a highly repressed state, exploratory experiments were carried out on the computer. It was essential to determine the minimum level of pool P_i in conditions where the system still remained operational, but in which growth would be extremely slow. After a series of studies, it was determined that when rate constant k_{18} was reduced to such a degree that its activity was only 5% of the original value (normal growth), the system was still operational but grew extremely slowly. A slight further reduction produced a completely dormant system, where entities of the system were kept on a constant level as a function of time. Further reduction of k_{18} produced a decaying system. These experiments are shown in Figs. 132 and 133. It is evident that

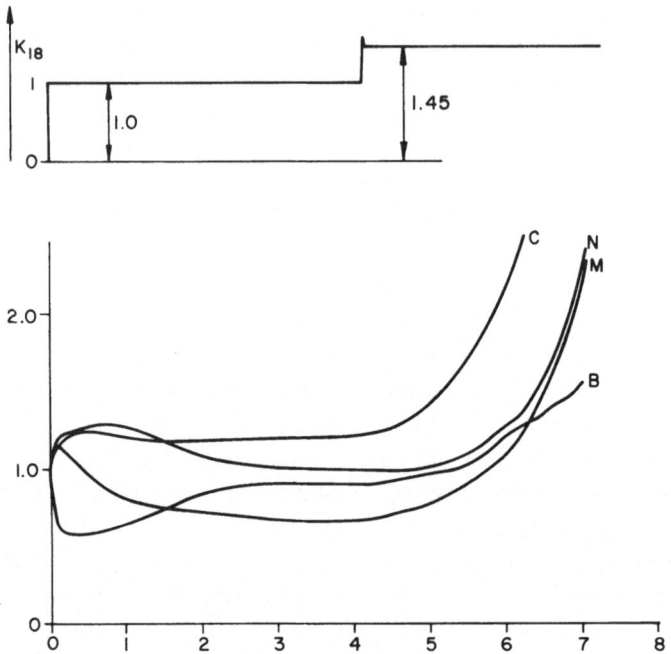

Fig. 132. The effect of transport enzyme E_t activity on the growth of the system. The value of the rate constant k_{18} is selected for very slow growth. At a certain time k_{18} is increased by 45%. The system starts to grow more rapidly. Template N, messenger M, transport RNA C, and ribosome B are recorded.

the system, after passing the initial transient, grows extremely slowly when compared to normal growth.   At such a level of injury or growth, the control system maintains itself rather indefinitely. When there is partial release of inhibition, for example, when k_{18} is increased 45% from this initial value, growth of the functional entities is resumed.   At the time indicated by the arrow, k_{18} was increased and it is evident that pool P_i increases rapidly initially, after which a steady growth follows.   Pools P_a and P_n both reveal a small initial transient but grow subsequently in a different fashion: While pool P_n grows rather smoothly, P_a reveals several slow transients.   Other components, also after the initial period, will grow steadily.   It is evident that the system is capable of organizing its increased growth from a rather static state when more pool is brought into the system. From the point of view of cellular dormancy, it is of interest to note that this system is capable of maintaining its functional organization and activity in conditions where the transport of the external pool is highly reduced. It is well known in cellular physiology that cells in many tissues have a slow replacement rate. In such conditions there is mainly maintenance synthesis instead of growth. However, growth can be initiated in such tissues, for example, when a tissue-specific hormone is introduced into the system. This suggests that a rapid and smooth growth activation occurs at a specific regulatory site within the cell. Such an experiment is simulated here, where a drastic change occurs at pool transport level; it is shown in Figs. 134 and 135, when the rate constant k_{18} is increased at a certain time from an initial value 1 to the value 3.2. Experiment reveals that a very rapid growth activation of the system is taking place. P_i, of course, is the first to increase, and this rapid increase again produces compensatory reduction in E. Other pools increase subsequently, and all components shortly begin to grow rapidly.

These studies on the model-system suggest that it is possible to exercise control of growth on the membrane trans-

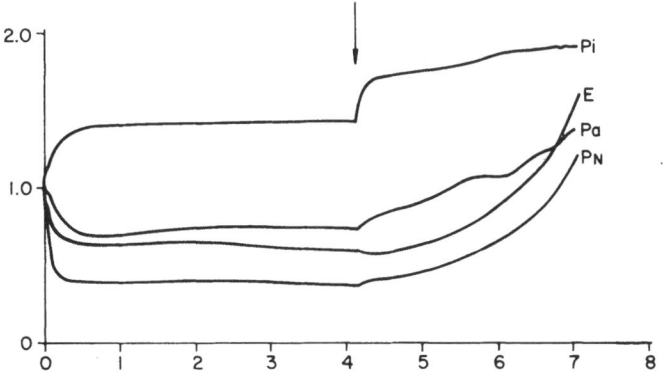

Fig. 133. Same experiment as Fig. 132. Pools P_i, P_a, and P_n and enzyme E are recorded.

port level. Here in the model-system we simulated the properties of external membrane. Of course, in a cell we have
multiple sets of internal membranes, and all of those membranes can be used to a certain degree for control of the
growth. The nuclear membrane in particular is in a strategic
position to control the synthesis of many entities.

The aspects of abnormal growth are also indicated in the
experiment, Figs. 134 and 135. It appears that if there is an
excessive activation of the transport system, abnormal growth
may develop where growth rate may be much higher than in
normal cells.

e. General Conclusions

The model-system studies have repeatedly demonstrated
that cellular growth can be abolished by inactivation of prac-

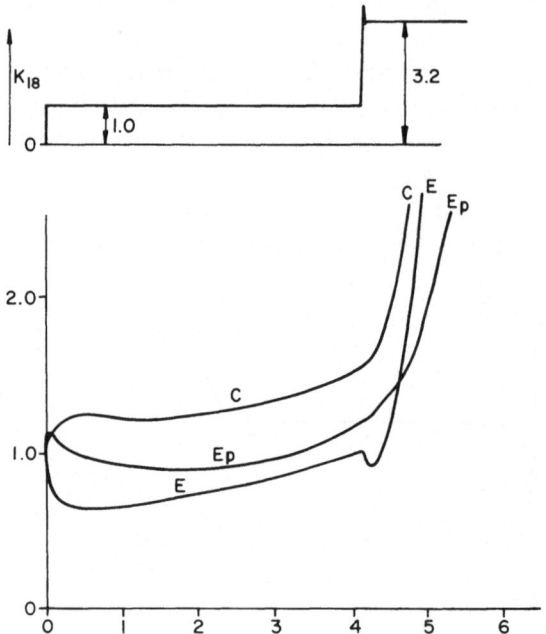

Fig. 134. The growth activation on enzyme E level. At a certain time the rate
constant k_{18} value is increased 3.2 fold, and rapid cell growth will follow. Enzymes
E and E_p and transport RNA C are recorded.

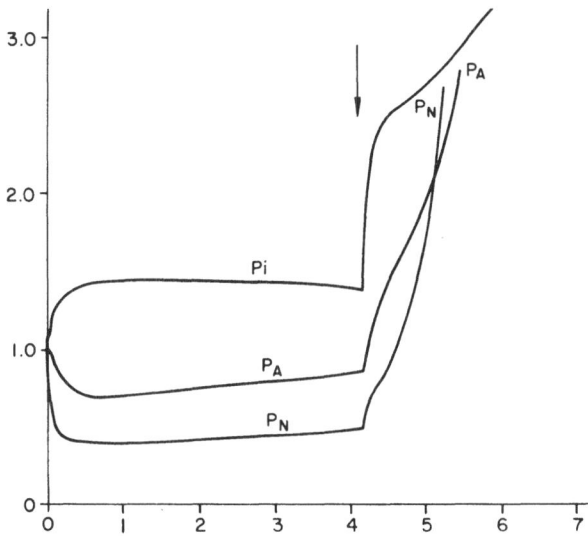

Fig. 135. Same experiments as Fig. 134.   Pools P_i, P_a, and P_n are recorded.

tically any major functional entity. The degree of damage
depends mainly on the extensiveness of the application of the
inactivating agent and on the duration of the application. As
was pointed out previously, complete and permanent inactivation of one essential gene is sufficient to destroy the functionality of the system, provided that this gene is really essential
and that its function could not be substituted by some alternative metabolic process. However, when gene inactivation is
only temporary, then the time factor, as well as the general
functional characteristics of the system, enters into play.
Complete genetic block, which is either a temporary or a
permanent partial block, represents injury which is conditional. In cellular biology the phenomenon that the cell
can be inactivated on the level of non-genetic functional
entities has not been as well recognized. We postulated over
ten years ago, on the basis of experimental evidence, that
there was operative in the cellular injury process a complex
type of interaction, which could not be interpreted in terms
of a simple target concept. At that time, it was intuitively
postulated that multiple pattern injury represents a mode of

disorganization of cellular growth and function [19, 20, 21, 22]. Experimental studies revealed that broad features of functional organization were effected. A number of quantitative measurements on the cellular processes [21, 22] suggested that there was no cleancut evidence that some well-defined injury was always the cause of the loss of cellular viability. Current model-system studies show that the functionality of the system may indeed be abolished by a process of progressive disorganization. Once extensive disorganization has occurred, then functional elements can no longer grow in a synchronized and balanced manner. Instead of a growth, there is a gradual decay.

These simulation experiments also provide a basic clue to cellular viability. The loss of an interaction pattern which leads to harmonized growth seems to be a general feature of the nonviable system. On the other hand, if one knows which elements cause the disorganization, one may reverse the trend of disorganization by applying proper counter-measures, and thus restore the functionality of the system. Furthermore, model-system studies permit one to project that cellular recovery is highly conditional. Model-system studies also suggest that if a cell is not capable of self-recovery, it can be helped toward recovery if proper external agents are applied (Figs. 107 to 114). It is indeed gratifying to us that computer experiments support the earlier conclusions which had been postulated on the basis of experimental work. At that time, we stated: "Experimental data suggests that in general a more effective cellular recovery from injury can take place when a combination of treatments are used for the restoration of viability. However, for effective treatment it is essential that the proper diagnosis in the terms of metabolic reaction and synthetic processes should be made for injured cells. Consequently proper metabolites, co-factors, and other specific means can be applied to obtain the conditions of optimum recovery." [22].

Growth control and growth stabilization experiments reveal that only on the level of certain functional entities is it possible to obtain a smooth, uniform control without destroying the balance of functional entities. Such features are essential for initiating cellular growth from a dormant state or suppressing the growth of a rapidly growing cell. The creation of static growth conditions in the model-system by the regulation of a limited number of functional entities suggests that growth can be suppressed by external agents in cellular systems. This concurs very much with experimental observations where some agents simply reduce the cell growth or division, while others destroy cell functional activities completely. For example, agents which limit bacterial growth are called bacteriostatic, and agents which kill the cell are bacteriocidal. This is an historical classification, but it is still currently used in literature. It has even been suggested that some types of cellular interactions with external agents lead to bacteriostatic effects, and others to bacteriocidal ones. Davis and Feingold [11] state: "The agents that damage wall, membrane, or DNA are bacteriocidal; the agents that are known to act by inhibiting enzymes (other than those specifically concerned with wall or DNA formation) have all been bacteriostatic. One is tempted to generalize that the bacteriocidal effects depend on the uniqueness of each part of the DNA (in a uninucleate cell) and of each part of cell envelope; irreversible damage to part of either cannot be compensated by the presence of intact material of the same class elsewhere in the cell." Model-system studies indicate that when severe damage occurs on the level of any entity, inactivation of the system results. Obviously, the living cell is much better regulated than a model-system. However, a basic concept which has been established from model-system studies reveals that disorganization of the system can be produced on the level of practically any functional entity, and extensive disorganization leads to decay of the system. Therefore, the postulation of Davis and Feingold that cellular inactivation occurs only when there is damage of cell membrane, cell wall, or DNA seems to be arbitrary. The concept of "bacteriocidal" means that, in the presence of the agent, the cell is not able to grow, since there is no synthesis for cellular growth and division. However, when this agent is removed, the cell may be capable of initiating growth again, but the cellular fate is conditional. When such an experiment is carried out in practice, the time factor, of course, will play an important role. If bacteriostatic conditions are continued for a long time, cellular destruction will take place, because there is no compensatory synthesis. Any biological system will decay by the inherent property that all functional entities are unstable, and in order to maintain the system maintenance synthesis should occur. On the other hand, for a limited amount of time, the cell can sustain a certain amount of decay and still be able to recover, when the condition for the growth is soon re-established.

At this point it would be worthwhile to make a few comments on cellular viability studies in a medium containing a mass population. Overall kinetics of cellular deaths of a mass population may appear simple, but it is an extremely complex phenomenon in terms of biological processes. We should like to make a few comments on cellular viability studies within a mass population on the basis of experience which we have obtained by analyzing the model-system. Model-system studies indicate that a viable system can sustain a certain amount of disturbance. The resulting disorganization depends upon how long the state of disturbance of normal functional processes is maintained, as well as on the stability of functional entities of the system. It is a well-known experimental observation that all cells do not have exactly equal characteristics. For example, if a group of synchronized cells are placed into growth medium, within a few generations synchrony is lost, indicating that all cells are not equal in terms of metabolic and synthetic processes. This is, of course, to be expected, since the cell is a highly complex organizational and functional system, where statistical events in terms of molecular interaction and reactions represent the basic processes. Consequently, both the structure and the function of entities are probabilistic in nature. Furthermore, variations on the level of the single entity are supplemented by sets of higher-order interactions. Thus, systems develop having a large number of probabilistic states. Consequently, a population contains species which have a wide variation in survival potential under normal conditions. Subsequently, when the cellular population is put under stress, one would expect that there would be among individual cells a wide range of deviations from the average behavior. If it is decided to destroy a cellular population completely—a process, for example, which is essential for sterilization—then we may be able to kill off the majority of population with a relatively small dose, but for the elimination of few individuals which have high survival potential (extremely effective functional organizations) a large dose may be required. We feel that, on the basis of general concepts developed in computer studies on the individual cell model-systems, new theories could be developed which could determine the kinetics characteristic of population death.

8. PRELIMINARY STUDIES ON GENETIC DIVISION AND ITS EFFECT ON GROWTH

Our model-system studies so far have been carried out on the growth processes without considering genetic or mitotic division. As was pointed out initially, in order to include those two cellular mechanisms, one has to develop a more advanced type of model-system. These will be discussed in the second part of this book, but this type of model-system cannot, at the present time, be analyzed on the computer. However, some limited studies on genetic division can be carried out on the present model-system. It is of special interest to see how the doubling of the number of genes affects the growth of various functional entities. For example, one may reason on the basis of straightforward linear thinking that doubling of the genes would also double the growth rate of the system. In order to simulate genetic division in our present model-system (Fig. 1), one has to supplement mathematical expressions (Table III) with additional programming features. For this purpose, an electronic switch which permitted doubling of gene concentration at any time of growth was introduced into the system. The

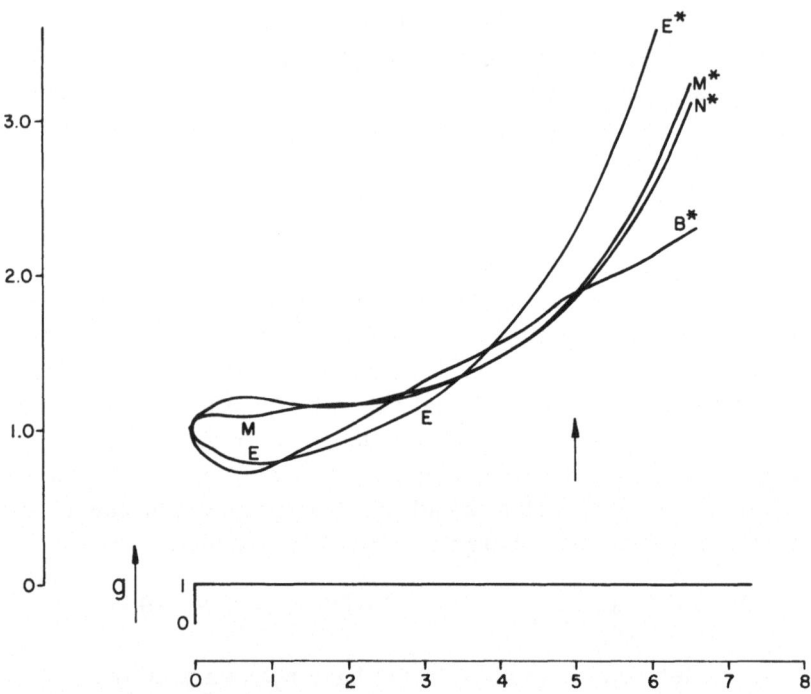

Fig. 136. The effect of genetic division on the growth of the system. This is a control showing growth in normal state without gene division. Enzyme E, messenger M, template N, and ribosome B are recorded. All elements marked with star are components growing without gene division.

following considerations have to be taken into account: Chromosomal division time varies in a wide range in various cells. There is no definite reference point for selecting a particular time.   Here we assume that two-thirds of a "generation time" was a convenient and reasonable point at which to start genetic division.   Since our interest is mainly phenomenological, genetic division is carried out as a step process. It could also be established as a gradual process, but details of genetic division are still too obscure to establish a more refined division mechanism.   For example, it is not known when genes become functional during the division—whether they become functional when the gene has divided, or whether they will become functional only when all of the chromosomes have divided. Since the basic issues are not clear in cellular biology, a single-step operation for division of all genes was considered to be a satisfactory approximation. Since division of the gene is carried out as a completely independent process, there is no utilization of pools. This is, of course, a cross-approximation, but since DNA represents such a small fraction of the total cellular mass, the amount of pools used for DNA synthesis would be comparatively small. Consequently, the effects of genetic division on the nucleotide pool can at present be ignored.

Figure 136 shows the normal growth of the principal entities of the system. In Figs. 137–139, the growth of some entity in normal conditions is occasionally represented in order

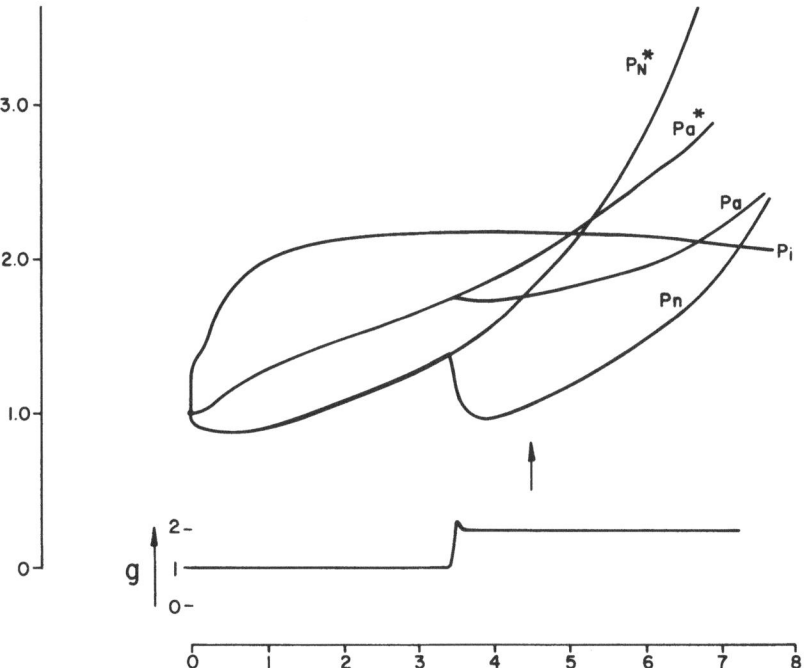

Fig. 137. The effect of gene division on the growth of the system. At the time indicated on the graph the value of all genes (G) are doubled. Pools P_i, P_a, and P_n are recorded. Starts indicate the normal components.

to compare systems with dividing and nondividing genes. Figure 137 shows the effect of genetic division on the pools. It is evident that after the division of genes, there is a high demand for P_n pool. This is to be expected, since gene activity has been doubled, resulting in additional RNA synthesis. However, pool P_n concentration, after passing through a minimum, starts to increase again. The growth rate of pool P_a is also reduced after genetic division, as indicated by the difference between P_a^* and P_a, but P_a gradually starts to build up again. There is also a reduction of pool P_i (not shown in Fig. 137). Figure 138 shows that E_p polymerase has a slight increase in concentration after the genetic division, but the growth rate is subsequently reduced and new growth is only slightly higher than normal growth (E_p^*). Messenger M_p and template N_p are growing at a slow rate, and the effect of

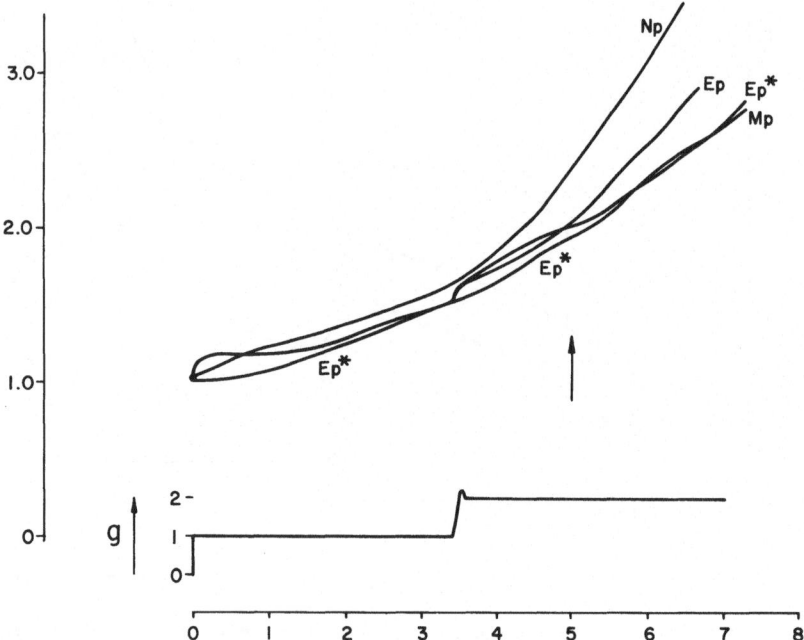

Fig. 138. Same experiment as in Fig. 137. Template N_p and enzyme E_p are recorded.

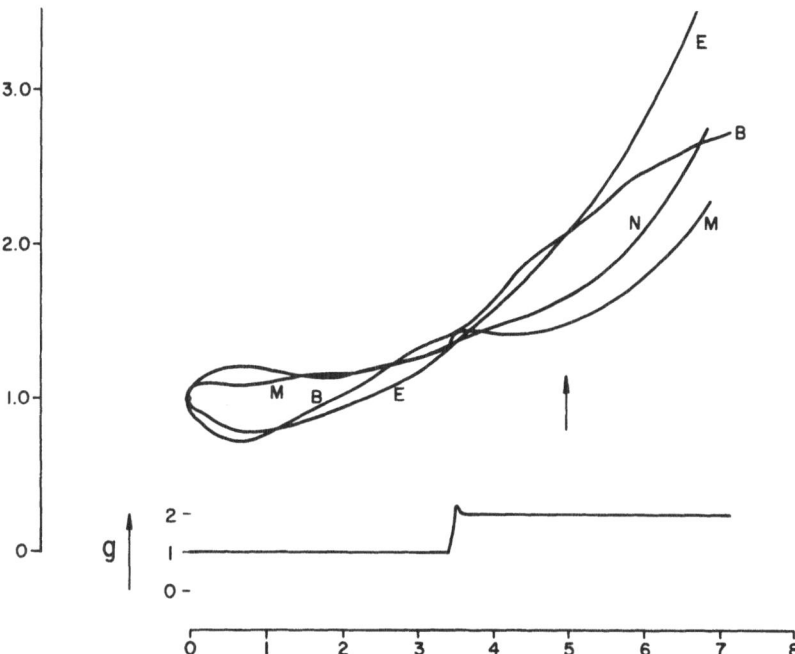

Fig. 139. Same experiment as in Fig. 137. Enzyme E, template N, messenger M, and ribosome B are recorded.

division does not seem significant. Figure 139 shows that after genetic division there is a drastic increase in messenger M and template N concentration. Subsequently, M concentration stays constant for a while and then starts to increase again. Ribosome B concentration is only slightly increased, while enzyme E growth increases markedly.

The following results have been obtained when measurements of normal growth and growth with gene division are compared at the end of a "generation time." The following are reduced: a) enzyme E, 9%; b) messenger M, 20%; c) template N, 11%; d) pool P_n 41%; and pool P_a, 20%. The following were increased: a) ribosome B, 7% and enzyme E_p, 5%. These results are surprising indeed. Instead of increasing, the gene concentration decreased the growth rate of the system. It appears that for optimum growth the system has to be properly

balanced.   Excess concentration of any particular entity,
deviating from optimum concentration, represents a reduction
in efficiency of growth. This may be so because a particular
element is not fully utilized.   However, these effects are
temporary, since if this gene duplication experiment is extended
beyond a generation time, the system gradually readjusts,
reaches a new balance, and starts to grow rapidly. Since the
rapid growth of the system occurs far beyond the generation
time, one can project some conclusions into the cellular
system.   The principal conclusion to be drawn from these
experiments is the demonstration that genetic division may
not materially affect cell growth shortly after division occurs.
There may be a slight, overall growth rate reduction during the
early phase at division, with growth occurring later. The final
result depends on the particular time during generation at
which the division takes place. This suggests that if a genetic
division is speeded up, the cell volume would gradually decrease
in successive generations until it established itself as a balanced
system at a reduced level. Consequently, the conversion of a
slowly growing cell into a rapidly dividing cell should reduce
the size of the cell. These conclusions are derived on the basis
of the condition that external pool P_e is constant. However,
if external pool P_e is increased, different results would be
obtained. For example, the increase of the external pool would
produce a more rapidly growing and larger cell. This could
lead effectively to a reduction of generation time. The effect
of genetic division on the growth rate reveals that the doubling
or tripling of some functional entity does not necessarily
increase the system's growth, and can in fact have the reverse
effect.

9. GENERAL COMMENTS FOR THE ANALYSIS OF
THE MODEL-SYSTEM

In summary, a functional model-system has been helpful in
elucidating the problem of cellular growth kinetics. Computer
studies with the model-system have made it possible for us to
obtain some information in regard to interrelationships between

various functional entities, as well as on the effect of various parameters on the growth process. These studies include the effects of the various entities on normal growth. The phenomenon of cellular viability and the recovery of viability has been elucidated, and it has been shown that the organizational pattern plays a dominant role in a functional system. Finally, genetic division has also given insight into its effect on the growth process.

While the simulation of the cellular system with the model has produced some highly interesting results, nevertheless, the complexity of the problem raises many new issues. In order to gain further insight into biological processes through model-system analysis, the model-system must be expanded. Above all, it should contain a greater amount of regulatory mechanisms. However, in our experience it was extremely difficult to obtain a solution on the computer for the differential equations representing the model-system. Sometimes it was in fact deemed impossible. Consequently, one has to introduce new features into the model-system gradually. However, since a functional system has been obtained, there is no apparent difficulty to supplementing, within certain limits, additional features involving regulatory mechanisms at various levels. This can be done with available computer facilities. However, more expanded model-systems will also require new developments in computer technology. Here again, there is no apparent limitation from the point of view of computer technology; the problem is mainly economic.

10. REQUIREMENTS FOR COMPUTER PERFORMANCE AND LAYOUT

This is a mathematical problem. The model-system (Fig. 1) is represented by nineteen simultaneous differential equations (Table III) containing thirty-three rate constants. In order that a functional system be established, the computer must have certain performance characteristics as well as a suitable physical layout. The organization of a functional system on the computer is time-consuming and tedious. It also makes great demands on the person who organizes the system, particularly in terms of memory, as well as evaluation and pertinent intellectual reasoning during the operation. Prolonged occupation with the organizing process is necessary for gaining experience and insight into the behavior of the system, when parametic adjustments are made. However, prolonged occupation with computer manipulation and accompanying mental efforts produce fatigue, and, therefore, finally counterbalance the positive gains. For a solution to be obtained to such a problem, it is essential for the computer to possess certain performance characteristics. We shall therefore outline a number of essential requirements on the basis of our own experience:

1. The computer must contain a sufficient number of operational elements (amplifiers, integrators, multipliers, electronic switches, potentiometers, etc.). These must perform with great dynamic accuray.

2. The programming patchboard must be removable so that the problem can be taken off the computer. This is imperative for prolonged studies as well as for organizing the system.

3. All operational control knobs and switches should be within easy hand-reach of the operator.

4. Computer performance should have the following principal characteristics: *

   a. High speed, a compressed time scale, a broad frequency band for accuracy and reproducibility.

   b. It must operate with high dynamic accuracy in real time, as well as in repetitive performance. The latter is required for organizing the system by using an oscilloscope for observations. If the repetitive solution on the scope is not accurate, it is not possible to organize the functional system. The real-time solution is used for recording the data with an X–Y recorder. Switching from repetitive to real-time and vice versa should be simple.

*Note: In these studies, the GPS Instrument Company (Newton, Massachusetts) General Purpose Computer, Model GPS10000, was used.

c. The computer should be able to maintain stable performance when square-wave potentials are introduced into the system, and it must be able to recover rapidly from overload conditions.

d. The noise level must be low. This is extremely important when characteristics of the system are studied in conditions of decay (for example, in studies of viability).

REFERENCES

1. F. Heinmets, J. Theor. Biol. 6: 60 (1964).
2. F. Heinmets, Electronic Aspects of Biochemistry (New York: Academic Press, 1964), p. 415.
3. O. Maaløe, and C.G. Kurland, Cell Growth and Cell Division (New York: Academic Press, 1963), p. 93.
4. O. Maaløe, The Bacteria, Vol. IV (New York: Academic Press, 1962), p. 1.
5. J.M. Mitchison, J. Cell. Comp. Physiol., 62: 1 (1963) J. (Suppl. 1, Part II).
6. J. Monod, and F. Jacob, Cold Spring Harbor Symp. Quant. Biol., 26: 389 (1961).
7. F. Jacob and J. Monod, J. Mol. Biol., 3: 318 (1961).
8. J.D. Watson, Science, 140: 17 (1963).
9. O. Greengard, and P. Feigelson, J. Biol. Chem., 236: 158 (1961).
10. W.E. Knox, Trans. N.Y. Acad. Sci., 25: 503 (1963).
11. B.D. Davis, and D.S. Feingold, The Bacteria, Vol. IV, (New York: Academic Press, 1962), p. 343.
12. D.E. Lea, Actions of Radiations on Living Cells (Cambridge: Univ. Press, 1947).
13. O. Rahn, Injury and Death of Bacteria by Chemical Agents (Normandy, Missouri: Biodynamica, 1945).
14. T.H. Wood, Advances in Biological and Medical Physics IV, (New York: Academic Press, 1953).
15. L.H. Gray, The Initial Effects of Ionizing Radiations on Cells, (New York: Academic Press, 1961), p. 21.
16. G.W. Barendsen, ibid., p. 183.
17. H. Marcovich, ibid., p. 173.
18. F. Heinmets, Physiological Concept of Cellular Injury and Death: AAS Symposium Berkeley, California, 1954.
19. F. Heinmets, Int. J. Rad. Biol., 2: 341 (1960).
20. F. Heinmets, J.J. Lehman, W.W. Taylor, R.H. Kathan, J. Bact., 67: 511 (1954).
21. F. Heinmets, and R.H. Kathan, Arch. Biochem. Biophys., 53: 205 (1954).
22. F. Heinmets, and J.J. Lehman, Arch. Biochem. Biophys., 59: 313 (1955).
23. R. H. Haynes, "Molecular Localization of Radiation Damage Relevant to Bacterial Inactivation," Physical Processes in Radiation Biology, (New York: Academic Press, 1964), p. 51.
24. R.B. Webb, Physical Processes in Radiation Biology, (New York: Academic Press, 1964), p. 267.
25. E.L. Powers, Phys. Med. Biol., 7: 3 (1962).

PART II

Descriptive Analysis of an Advanced Cellular Model-System

1. INTRODUCTION

Quantitative analysis of the first model-system (Fig. 1) yielded operational information in regard to major functional entities in the process of growth. It also revealed that the functional organization of the system was dependent not only on concentration levels, but also on concentration patterns of the functional entities. Excessive modification of these conditions could lead to the loss of a functional system.

In order to extend the studies of growth phenomenon from the single-cell level to the tissue or organism level, it is essential that a model-system also contain features of genetic and mitotic division. Furthermore, interactions between individual cells, as well as between various tissues and organs which lead to an integral system, are extremely important in the analysis of a multicellular organism. These cannot, at the present level of the "art," be included in the model-system. However, after the descriptive analysis of a single-cell model-system, a speculative discussion on abnormal growth development on the tissue interaction level will be carried out.

Modification of the functional properties and the growth rate of individual cell(s) can cause changes within the tissue function. This, in turn, can affect the total behavior of the organism. Uncontrolled or excessive cell growth may lead to the conditions of malignance, while excessively retarded growth may slow cellular replacement and may lead to degeneracy. Change of a cellular functional process, for example, such as production of hormones by a specific tissue, can progressively affect the total system. In this section of the book we shall analyze an advanced model-system, and shall subsequently carry out several analyses of certain phenomena resulting from cellular interactions. Particular attention will be paid to the mechanisms and modes of interaction which lead to loss of regulatory controls and the development of abnormal growth.

In Part I of this book we carried out studies on quantitative interrelationships between various functional entities in diverse conditions of growth. Furthermore, we experimented to determine at which particular sites of the model-system it was possible to exercise effective overall growth control. The results suggest that only a limited number of functional entities can be utilized for the stabilization of growth, and that these particular entities could be made operational by the action of regulatory mechanisms. In order to study cellular growth, and regulatory and trigger mechanisms in various conditions, these features were integrated into the advanced model-system (Fig. 140). It was of special interest to study the possibilities of abnormal growth development on the model-system basis. This approach was considered to be worthwhile, since the enormous experimental data found in the literature are extremely amorphous and do not permit one to establish any uniform concepts for the interpretation of abnormal growth processes [36, 56–58]. At the present level of information, any attempt to achieve a unifying concept of abnormal growth using only experimental data is a fruitless effort. Therefore, we decided to use an alternative approach, in which a model-system analysis is carried out on the basis of a broad background of experimental information, without attempting to justify every conclusion with a specific experimental fact. Rather, we attempted to analyze the subject from a theoretical point of view and to speculate freely on various possible mechanisms which could be operational in abnormal growth conditions. However, this speculative approach is still confined to restraints

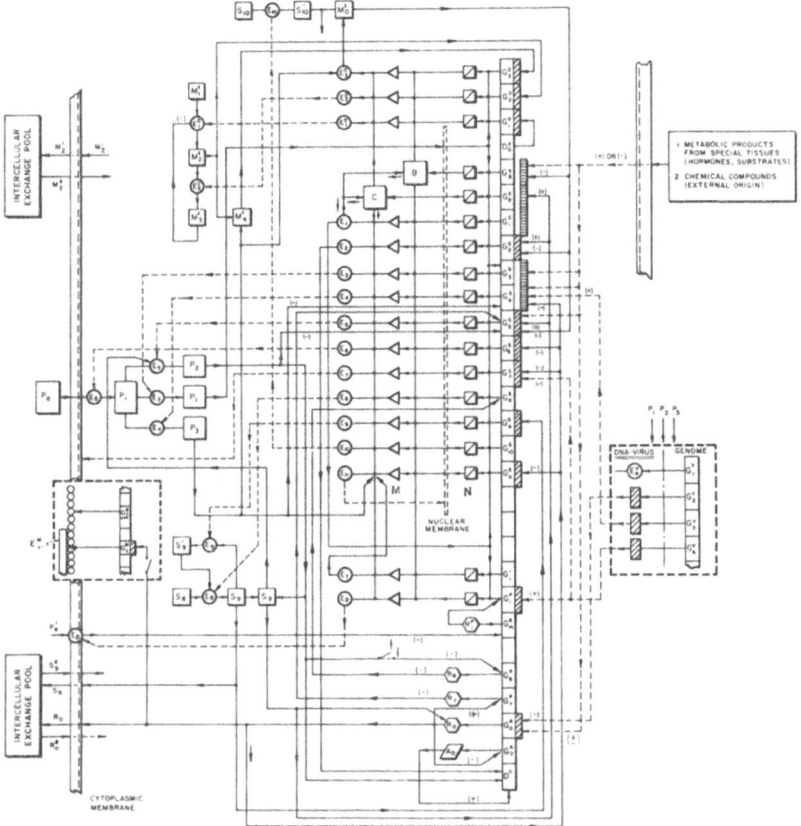

Fig. 140. Advanced model-system for analysis of cellular growth and division processes.

of general factual information on growth phenomena. This way it is possible to raise significant questions and to suggest some potential areas of research. It is felt that the phenomenon of abnormal growth is a problem which is going to be with mankind for an indefinite time and needs exhaustive study over a wide area of cellular biology. It appears that a speculative analysis carried out on a model-system would be instructive, especially since the subject could be discussed in a logical manner without specific restraints. Any attempt to correlate heterogeneous experimental data derived from various unrelated experiments (performed in different species in a variety of experimental conditions) is not, in our opinion, a profitable task.

However, we shall attempt to analyze some specific experimental information in relation to abnormal growth. In order to separate the abstract model-system analysis from factual analysis, the latter will be treated in a separate section of the book (Part III).

2. MODEL-SYSTEM FORMULATION

a. Introduction

The basic functional entities utilized in the advanced model-system are for the most part the same basic entities which were presented in the previous model-system (Fig. 1). However, some new entities are added here for formulating regulatory and division processes. There is a larger number of functional gene groups, and genes have different states of activity. The general scheme of the advanced model-system is represented in Fig. 140.

b. Terminology and Symbols

1. Pools for Nutrition and Synthesis

   P_e — Extracellular nutrient pool

   P_i — General intracellular metabolic pool

   P_1 — Specific pool for RNA synthesis

   P_2 — Specific pool for DNA synthesis

   P_3 — Specific pool for protein synthesis

   P'_e — Pool derived from proteolytic activity on adjacent cells (for the analysis of abnormal growth)

2. Entities which Participate in the Regulatory Processes

   R_0^e — External DNA division repressor

   S_9^e — External DNA division inducer

   $S_0...S_n$ — Metabolites participating in triggering DNA division

M'_2 — Internal inducer for mitosis

M_2^e — External inducer for mitosis

$M_0^1...M_n^1$ — Metabolites participating in the process of mitosis

3. Enzymes (Groups)

E_1 — RNA polymerase

E_2 — DNA polymerase

E_3 — Aggregate system of enzymes which convert part of P_i into P_1

E_4 — Aggregate system of enzymes which convert part of P_i into P_3

E_5 — Aggregate system of enzymes which convert part of P_i into P_2

E_6 — Aggregate system of enzymes which convert P_e into P_i

E_7 — Controls transport properties of cell membrane

E_8, E_9 — Participate in trigger mechanism of DNA division processes

$\left.\begin{array}{l} E_1^D, E_2^D \\ E_3^D, E_{10} \end{array}\right\}$ — Participate in process of mitosis

E_{11} — Controls transport properties of nuclear membrane

E_t — Amino acid transport enzyme

E_1^M — Controls formation of external membrane

E_p — Proteolytic enzyme acting on adjacent cells

E_p^v — Viral polymerase for DNA synthesis

4. RNA Components

B — Ribosomal

C — Transfer

M (1...n) — Messengers

N (1...n) — Templates

$R_0...R_n$ — Repressors affecting gene activity

A_0 — Activates a genetic site for DNA division mechanism

5. Genes

$$\left.\begin{array}{l} G_1^D ...G_n^D \\ G_1^S ...G_n^S \\ G_1^P \end{array}\right\}$$ — Involved in the synthesis of respective messenger RNA's

$G_0^R ...G_n^R$ — Involved in the synthesis of respective repressors ($R_0...R_n$)

G_0^A — Involved in synthesis

D_0 — Active starting point for DNA division

D_0^X — Terminating point of DNA division

G_B^S — Ribosomal RNA (B) synthesis

G_C^S — Transport RNA (C) synthesis

G_1^t — Involved in synthesis of messengers for the amino acid transport enzymes

$G_1^V ...G_n^V$ — Viral genetic elements

c. States of Functional Activity of Genes

A normal ground-state gene has value of unity. However, gene activity can be modified when it interacts and forms a complex with some activator or deactivator compound (i.e., hormones). The following designation will be used:

		Activity Coef. for Gene
1.	$\overset{\circ}{G}$ — Ground state	k_1
2.	$\overset{*}{G}$ — Activated state	k'_1
3.	$\underset{*}{G}$ — Deactivated state	k''_1

where: $k'_1 > k_1 > k''_1$. Basic premise: Once a gene has been activated or deactivated by a compound, gene activity cannot be further increased or decreased by some additional compound, until it has returned to the ground state.

3. THE GROWTH PROCESS AND ENZYME SYNTHESIS

Basically cellular growth represents an increase in the number of functional elements per unit cell. Cell growth is terminated by division, wherein the average number of functional units is halved. However, there may be occasions where the cell does not divide, when the mass has doubled, leading to an enlarged cell. Before analyzing the integrated growth pattern of the functional elements, the synthesis of major functional elements will be described.

A detailed kinetic analysis of induced enzyme synthesis has been presented by us elsewhere [34, 35]. However, in this system a simultaneous synthesis of many enzymes occurs, in which several other functional elements participate, and, consequently, it is essential to carry out an extended formulation of the process here.

In Fig. 140, the basic schemes for enzyme synthesis are included. The following elements participate in the process: ribosome B, transfer RNA C, messenger RNA M, templates, and pools P_1 and P_3. Since all enzymes are considered to be synthesized by a similar process, only one case will be analyzed here. Flow equations for enzyme and RNA synthesis are presented in Table IV.

Steps 1 and 2 represent a condensed version of ribosome synthesis. Gene G_B^S interacts with polymerase E_1 forming a complex. The latter interacts with RNA pool P_1, yielding ribosomal RNA M_B, which is converted subsequently into ribosome. Ribosomes are relatively stable, but they gradually decay into general pool P_i.

In step 3, ribosomes interact with polymerase E_1 and form a reversible complex. The latter may be transformed in a subsequent step into a nonactive form $[BE_1]^*$, which can be activated with a specific inducer substance (n_1). This process will be analyzed in relation to growth in another section of the book. Gene G_C^S interacts with E_1 (step 5), yielding a complex

TABLE IV

Formulation of Enzyme Synthesis

1. $G_B^S + E_1 \rightarrow [G_B^S E_1] + P_1 \rightarrow M_B + G_B^S$
2. $M_B \rightarrow B \rightarrow P_i$
3. $B + E_1 \rightleftharpoons [BE_1] \rightarrow [BE_1]^*$
4. $[BE_1]^* + n_1 \rightarrow [BE_1]$
5. $G_C^S + E_1 \rightarrow [G_C^S E_1] + P_1 \rightarrow M_C + G_C^S$
6. $M_C \rightarrow C \rightarrow P_i$
7. $C + E_1 \rightleftharpoons [CE_1] \rightarrow [CE_1]^*$
8. $[CE_1]^* + n_1 \rightarrow [CE_1]$
9. $G_I^S + E_1 \rightarrow [G_I^S E_1] + P_1 \rightarrow M_1 + G_I^S$
10. $M_1 \rightarrow P_1$
11. $M_1 + B \rightarrow N_1$
12. $E_t + P_3 \rightarrow [E_t P_3] + C \rightarrow [CP_3] + E_t$
13. $[N_1] + [CP_3] \rightarrow [N_1] [CP_3] \rightarrow E_1 + M_1 + B$
14. $E_1 \rightarrow P_i$
15. $G_B^S + M_0 \rightarrow G_{*B}^S \rightarrow G_B^S + X$
16. $G_B^S + S_0 \rightarrow G_B^S \rightarrow G_B^S + X$

Additional conditions:

1. $[G_B^S E_1] + P \xrightarrow{k_1'} M_B + G_B^S$
2. $[G_{*B}^S E_1] + P \xrightarrow{k_1'} M_B + G_{*B}^S$
3. $[G_{*B}^S E_1] + P \xrightarrow{k_1''} M_B + G_{*B}^S$

where: $k_1' > k_1 > k_1''$

In example $k_1 = 1$; $k_1' = 5$; $k_1' = .1$

$[G_C^S]$. The latter interacts with pool P_1, producing an RNA
fraction, M_C, which subsequently is converted to transfer
RNA C.   It also decays into general pool P_i. Steps 7 and 8
represent a regulatory interaction between C and E_1.

In step 9, functional gene G_1^S and E_1 form a complex which,
after interaction with pool P_1, yields a gene-specific messenger
M_1.   The messenger is unstable and decomposes into RNA
pool P_1. Active template N_1, is formed when messenger M_1 and
ribosome B form a complex (step 11).   Amino acid transport
enzyme E_t is produced via gene G_1^t.   In step 12, amino acid
pool P_3 and E_t yield a complex $[E_t P_3]$, which subsequently
interacts with C, yielding free enzyme E_t and complex $[CP_3]$.
Template N_1 and complex $[CP_3]$ interact, yielding enzyme E_1.
Enzyme E_1 decomposes into internal pool P_i (step 14). It is
considered that all enzymes are formed by the same general
mechanism, in which the processes are initiated by the
respective genes.   It is also considered that repressors and
activators are derivatives of gene products, but the pathways
leading to their formation are unknown.

4. DIVISION OF GENES AND TRIGGER MECHANISMS FOR THE PROCESS

The mechanism of division of genes is considered here
only in a very general manner.   The process is duplication
of genes and the initiation of division, where timing and
control are related to other functional processes.   Genetic
division is considered to be rather weakly coupled with
mitotic process.   As a consequence, growth may proceed in
the absence of genetic division.   In cellular growth in these
conditions large cells would be produced, a phenomenon often
observed.   Normally, however, cells divide in constant growth
conditions at a definite time interval, yielding normal-sized
cells.   This suggests that mechanisms which initiate genetic
division have the characteristics of a trigger, but that the
trigger action is not compulsive.   It can be considered that a
metabolic signal is provided at a specific metabolic state to

initiate the DNA synthesis. However, this signal produces actual triggering only when metabolic conditions are proper for the DNA synthesis. Otherwise, the signal is "passed over" without the initiation of synthesis. This triggering process is repeated at the next division cycle, and then synthesis of genes may take place.

Since the division of genes represents an integral part of the growth of tissues and organs, it is essential that, in a model-system, the trigger mechanisms be included in order to simulate cell growth and division. Unfortunately, there is barely any experimental evidence on how these trigger mechanisms operate (1-7), and most material presented is speculative. It seems to be justified that an integrated operational model-system which deals with sequential events of the growth process should contain metabolic trigger mechanisms, even if they are only speculative. This is especially so for the analysis of cyclic processes, where the system has to have all essential mechanisms required for the organized performance and where trigger mechanisms are coupled with other metabolic and synthetic events. With this in mind, two metabolic trigger mechanisms were devised, one for genetic division and another for mitotic division. The basic point in genetic division seems to be that division is to be initiated at a specific locus on the chromosome, and that activation of the site must occur before division can be initiated.

A flow scheme for genetic division is presented in Table V.

The first step represents the activation of division starting site, D_0, by forming a complex with activator A_0. A_0 is available only under certain conditions and for a relatively short time. The basic requirement for initiating the mechanism of the division is considered to be the condition that pool P_2 for DNA synthesis has to be completed, and the last pool metabolite will initiate the events. At first, we shall describe the basic processes which are carried out by individual units and, subsequently, a sequential trigger mechanism will be analyzed. It is assumed that the activator is a product of a gene (direct or indirect). Here, we select for simplicity a

TABLE V

Gene Division and Trigger Mechanisms

A. Trigger mechanisms for the initiation of genetic division

1. $D^0 + A_0 \rightarrow [D^0 A_0] \rightarrow D^0$

2. $G_0^A + E_1 \rightarrow [G_0^A E_1] + P_i \rightarrow A_0 + G_0^A$

3. $G_0^A + R_0 \rightleftharpoons [G_0^A R_0]$

4. $A_0 \rightarrow P_i$

5. $G_0^R + E_1 \rightarrow [G_0^R E_1] + P_i \rightarrow R_0 + G_0^R$

6. $R_0 \rightarrow P_i$

7. $R_0 + S_0 \rightarrow [R_0 S_0] \rightarrow P_i$

8. $G_8^R + E_1 \rightarrow [G_8^R E_1] + P_i \rightarrow R_8 + G_8^R$

9. $R_8 \rightarrow P_i$

10. $G_8^R + P_2 \rightleftharpoons [G_8^R P_2]$

11. $G_5^S + P_2 \rightleftharpoons [G_5^S P_2]$

12. $G_5^S \rightarrow ---- \rightarrow E_5$

13. $G_8^S \rightarrow ---- \rightarrow E_8$

14. $G_9^{*S} \rightarrow ---- \rightarrow E_9$

15. $G_8^S + R_8 \rightleftharpoons [G_8^S R_8]$

16. $E_8 + S_8 \rightarrow E_8 + S_9$

17. $E_8 + S_9 \rightleftharpoons [E_8 S'_9]$

18. $E_9 + S'_9 \rightarrow E_9 + S'_9$

19. $S_9 + P_2 \rightarrow S_0$

20. $S_0 + E_5 \rightarrow \overset{*}{E}_5$

21. $S_9 + G_9^S \rightarrow \overset{*}{G}_9^S \rightarrow G_9^S$

22. $S_0 + G_4^{*S} \rightarrow \overset{*}{G}_4^S \rightarrow G_4^{*S}$

23. $S_0 + G_3^{*S} \rightarrow \overset{*}{G}_3^S \rightarrow G_3^S$

24. $S_0 + [G_C^S, G_B^S] \rightarrow \overset{*}{G}_C^S, \overset{*}{G}_B^S$

TABLE V (continued)

B. Synthesis of Genes

1. $P'_2 \rightarrow P_2$

2. $[D_0A^0] + E_2 \rightarrow [D_0A^0E_2] + P_2 \rightarrow 2D_0 + [E_2G_0^A]$

3. $[E_2G_0^A] + P_2 \rightarrow 2\ G_0^A + [E_2G_0^R]$

4. $[E_2G_0^R] + P_2 \rightarrow 2G_0^R + [E_2G_7^R]$

----

5. $[E_2G_3^S] + P_2 \rightarrow 2G_3^S + [E_2D_0^X]$

6. $[E_2D_0^X] + P_2 \rightarrow 2D_0^X + [E_2\overset{*}{G}_1^D]$

7. $[E_2G_1^D] + P_2 \rightarrow 2\overset{*}{G}_1^D + [E_2G_2^D]$

8. $[E_2G_2^D] + P_2 \rightarrow 2\overset{*}{G}_2^D + [E_2G_3^D]$

9. $[E_2G_3^D] + P_2 \rightarrow 2\overset{*}{G}_3^D + E_2$

direct mechanism, and step 2 shows the formation of activator A_0 by gene G_0^A. The active complex $[D^0A_0]$ has a limited life-time, and it decays back to inactive state D_0.

Step 3 indicates the repression of gene G_0^A by repressor R_0. The activator, A_0, is formed only when gene G_0^A is in the free state. Gene G_8^R produces repressor R_8 (step 8), and later represses gene G_8^S (step 15). R_0 acts as activator for gene G_7^R, which produces repressor R_7, which, in turn, partially controls gene G_5^S activity.

Genes G_5^S, G_8^S, and G_9^S produce respective enzymes (steps 12, 13, 14) by processes described in Table IV. Step 11 reveals that pool P_2 concentration is controlled at the genetic level. Functional units A_0, R_0, and R_8 decay to internal pool P_i (steps 4, 6, and 9).

Basically, one can consider that the trigger system is metabolically inactive during the greater part of the growth cycle. Activation of the system takes place when pool P_2 becomes functionally complete. The first event is the repression of G_8^R by pool P_2 (step 10). This terminates repressor R_8 formation and the release of gene G_8^S (step 15) repression. Synthesis of enzyme E_8 (step 13) follows. The internal metabolic pool contains a substrate S_8 which is converted to S_9 by enzyme E_8 (step 16). Gene G_9^S is in the inactive state and is activated by substrate S_9 (step 21). An enzyme, E_9, is synthesized by active gene G_9^S (step 14). Substrate S_9 is converted by E_9 into inhibitor S'_9 (step 18) and is suppressed when S'_9 interacts with E_8 (step 17), forming a complex $[E_8 S'_9]$. Consequently, there is high S_9 concentration only for the duration, while enzyme E_9 is not synthesized. When constant concentration of enzyme E_9 is established, S_9 concentration is highly reduced, since S_9 is considered to be relatively unstable, while complex $[S'_9 E_8]$ is stable. As a consequence, we have only a transient concentration of S_9. The following events will take place during this time. DNA pool P_2 and substrate S_9 interaction produces a compound S_0 (step 19), which has several functions. It activates enzyme system E_5 and thus increases the rate of pool P_2 formation. In addition, S_0 rapidly inactivates the repressor R_0 (step 7). Consequently, R_0 concentration is highly reduced, gene G_5^S is partially activated (reduction of R_7), and G_0^A is released from repression (step 3). Synthesis of division activator A_0 takes place (step 2), and site D^0 will be activated $[D^0 A_0]$. Synthesis of genes is initiated and proceeds sequentially (Section B, Table V). Since the complex $[D^0 A_0]$ has a limited lifetime, it returns to the normal inactive state, D^0. Further activation cannot take place until a new cycle is initiated, since S_0 disappears after a transient buildup. R_0 concentration increases again to the normal value, and gene G_0^A will be repressed (step 3). Growth of various elements could be increased by temporary gene activations (steps 22, 23, and 24).

Genes which participate in the cellular division mechanism are located at the terminal end of the chromosome. The duplication process of these genes is a part of the triggering mechanism of cytoplasmic division. Step 5 represents the duplication of gene G_3^S, and step 6 produces the formation as well as the activation of gene G_1^D. This later activates the trigger mechanism for cytoplasmic division.

5. CELL DIVISION

The cellular division mechanism is initiated before genetic synthesis is completed. The trigger mechanism represents a series of activation and inactivation processes. The principal steps are represented in Table VI. Activation of gene G_1^D leads to enzyme E_1^D formation (step 1). A metabolite M_1^1, already present in the cell, is converted by enzyme E_1^D into M_2^1 (step 2). It activates gene D_2^D (step 4), and enzyme E_2^D is formed (step 5). This enzyme converts M_2^1 into M_3^1 (step 6). Enzyme E_1^D is inactivated by M_3^1 (step 8), and M_2^1 concentration is highly reduced. Thus, we have a transient formation of M_2^1. Later it interacts with pool P_3 (step 9), forming M_4^1, and gene G_3^D is activated (step 10). Enzyme E_3^D is formed (step 11), and part of pool P_3 is converted into M_0^1 (step 12). This metabolite is involved in the cellular division process. Substrate S_{10} is converted to S'_{10} by enzyme E_{10} (step 14), followed by a complexing reaction between M_0^1 and S'_{10} (step 15). This complex leads, subsequently, to a polymerization process which can be visualized as leading to septum formation. The division trigger mechanism will be "closed down" when M_3 accumulates to a certain critical value and inactivates several operational division genes (steps 16, 17, and 18).

Let us consider the operational behavior of the cell division mechanism in conditions where there is interference by some external or internal agent during the process of division. Since enzyme E_{10} represents symbolically the enzyme system which will be operative in the division process, one can consider the situation in which some internal or external factor may affect this system, so that final metabolite (S'_{10}) will not be available after the trigger mechanism has been initiated at D_0^X level. What will be the outcome of the whole operation? It seems that the sequence of reactions and enzyme synthesis will proceed normally, except that the complexing reaction

TABLE VI

Processes in Cell Division

1. $\overset{*}{G}_1^D \; ---- \to E_1^D$

2. $M_1^1 + E_1^D \to M_2^1 + E_1^D$

3. $E_1^D \to P_i$

4. $G_2^D + M_2^1 \to \overset{*}{G}_2^D$

5. $\overset{*}{G}_2^D \; ------ E_2^D$

6. $M_2^1 + E_2^D \to M_3^1 + E_2^D$

7. $E_2^D \to P_i$

8. $M_3^1 + E_1^D \to [M_3^1 E_1^D] \to P_i$

9. $M_2^1 + P_3 \to M_4^1$

10. $G_3^D + M_4 \to \overset{*}{G}_3^D \to G_{3*}^D$

11. $\overset{*}{G}_3^D \; ----- \to E_3^D$

12. $E_3^D + P_3 \to M_0^1 + E_3^D$

13. $E_3^D \to P_i$

14. $S_{10}' + E_{10} \to S_{10}' \to P_i$

15. $S_{10}' + M_0^1 \to [S_{10}' M_0]$

16. $M_3^1 + G_2^S \to G_{2*}^S$

17. $M_3^1 + G_5^S \to G_{5*}^S$

18. $M_3^1 + G_1^D \to G_{1*}^D$

19. $M_1^1 \; ---- M_n^1 \to P_i$

(step 15) does not occur. Consequently, M_0^1 concentration is built up rapidly and inactivation of the genes (steps 16, 17, and 18) will occur immediately, without cellular division. Such an event simulates conditions where a binuclear cell is formed, since chromosomal division has taken place without cytoplasmic division. Of course, during the next division cycle, if system E_{10} is again not operative, a multinuclear cell is formed. Such a cell may divide in suitable conditions when the required trigger metabolite is supplied or built up by some compensatory mechanisms. Thus, a loose "metabolic impulse" type coupling between chromosomal and cytoplasmic division processes permits partial autonomy of each process. In biological systems, formation of multinuclear cells is a common phenomenon, especially in rapidly growing tissues.

6. "RESTING" STATE OF CELL AND RE-INITIATION OF THE GROWTH

After completion of the development of tissues and organs, the cell multiplication rate is highly reduced. In some tissues, cellular replacement can be fairly fast, while in others there is no further division (for example, in neural cells). It is obvious that cells which divide slowly or divide not at all have to have some residual synthesis for the maintenance of the system, However, the level of the synthesis can be extremely low when compared to actively growing tissue cells. Many dormant cells can be activated, and a rapid growth will follow when suitable inducers (i.e., hormones) are introduced into the system. We have previously analyzed this problem in a quantitative manner, using a simplified model-system (Fig. 1.) It is of interest to study this problem in a descriptive manner using a more advanced model-system. The question is: How does the cell acquire such dormant characteristics, and what regulatory elements participate to restore functional activity?

A speculative analysis will be carried out within the framework of the general model-system (Fig. 140). In an isolated cell, the rate of growth and division is internally controlled by various sets of regulatory mechanisms. However, in tissues and organs, adjacent cells exert strong influence on each other, and, consequently, independence of the growth process ceases to exist. It appears that because of a transfer and exchange of intracellular growth regulators and growth inducers, the cell loses its autonomy and acquires the characteristics of the communal specie. Since a single cell possesses functional mechanisms for growth regulation, and adjacent cells can influence the operational behavior of the functional elements, it is desirable to carry out the basic analysis of cellular interaction on a single-cell model-system. This can be carried out on the assumption that intracellular growth factors can be transported through the cell membrane, and this is facilitated by cell contacts.

For the analysis, we shall follow the pathways represented in Fig. 140. One can consider that the following cellular events would have significant bearing on the growth and division processes: a) change of transport properties of nuclear and cytoplasmic membranes; b) accumulation of intracellular repressors; c) inhibition of enzyme synthesis and conversion of templates into the inactive state. Growth-limiting conditions for tissue development could be reached when the principal repressor (R_0) concentration reached a sufficiently high value. At a high level of R_0, genetic division is terminated. R_0 concentration depends on intracellular synthesis and on an exchange of repressor between neighboring cells. When R_0 has reached at a certain value, the following events on the model-system can be considered to take place: Functional activity of gene G_{11}^S is highly reduced by interaction with R_0. As a consequence, concentration of enzyme E_{11} is highly reduced. Since E_{11} represents a part of the active transport system of the nuclear membrane, the interchange of metabolic elements between the nucleus and cytoplasm is reduced. In a similar manner, gene G_7^S is repressed, and the cytoplasmic membrane transport system becomes less active. Consequently, the transfer of external pool P_e into the intracellular space is also reduced. Therefore, pools P_i and P_3 are lowered, and the following regulatory events take place: In the free state, transport RNA C forms a complex with respective amino acids. This can be represented in general (see equation 12, Table IV) as

$$P_3 + C \rightleftharpoons [P_3C]$$

C is also complexing with RNA polymerase E_1, and thus another reversible complex present is

$$C + E_1 \rightleftharpoons [CE_1]$$

Since the formation of the first complex is much faster, only a minor amount of functionally nonactive $[CE_1]$ complex is present in the growing cell.   However, when P_3 is highly reduced, $[CE_1]$ complex becomes dominant and will be converted slowly into a completely inactive form, $[CE_1]^*$. It can be activated only by some activator or inducer to make it functional again (equations 7 and 8, Table IV). As a consequence, cellular synthetic processes are highly reduced, and only low-level "maintenance synthesis" occurs.   Tissues and organs in such a resting state carry out their basic functional processes. When there is a need to increase the function or size of the tissues, it is essential that an activator or inducer be introduced into the system.   As a consequence, the template system will be activated (equations 4 and 8, Table IV) and enzyme synthesis is resumed.   Cellular growth and division will follow.   In this type of activation process, there is no need for initial messenger RNA synthesis.   Of course, there are many tissue or organ-specific induction processes where the primary step may be RNA synthesis followed by protein synthesis and general growth. It is obvious that in such cases, many genes have to be activated simultaneously to produce all the RNA fractions necessary for the enzyme synthesis (i.e., C, B, and M). In all cases, it seems essential that some amounts of polymerases and transport enzymes be present for the initiation of synthesis. Of course, many enzymes may also be present in the system in the inactive state and may thus have to be initially activated.

In the present scheme, protein synthesis and amino acid pool P_3 play the principal roles in the growth process. Conclusions reached in quantitative analysis of growth regulation (Part I) are in agreement with this, since protein synthesis exerts an effective and smooth control on the cellular growth process.   Direct activation of genes by P_3 may be essential, but this mechanism is not clear (8).   It is obvious that an increase of external pool P_e could lead to an increase in growth, unless internal or external regulatory mechanisms suppress its effects. Transport mechanisms via cytoplasmic and nuclear membranes can effectively control the flow of metabolites and nutrients, and consequently have growth-rate determining properties. Activation transport mechanisms by hormones or substrates would gradually lead to an increase in growth rate. Such a growth may be only temporary, lasting only as long as activators are available. Growth will cease gradually when the activator is removed.

Any external metabolite or molecule which can inactivate repressor R_0 directly (equation 7, Table V) or indirectly will reduce the repression and lead to increased growth and division. For example, when S_0 is supplied by an external source or by adjacent cells, genes G_5^S, G_2^S, G_1^S, G_C^S, and G_B^S would be activated and synthesis occurs. Obviously, introduction of R_0 from an external source would lead to repression of growth. Furthermore, an increase of R_0 may initiate mechanisms whereby transport properties of the cytoplasmic membrane could be affected. This may lead to a further increase in internal repressor concentration. Permeability reduction could occur via activation of gene G_1^M (Fig. 140), which would lead to the formation of supplementary membrane components (E_1^M). Such a mechanism could terminate cellular growth when additional membrane components are formed enclosing the whole system.

A different type of growth phenomenon occurs when the internal cell division mechanism fails to operate (Table VI) after each chromosomal division. As a consequence, multinuclear cells could be produced. Introduction of activator M_2^c would initiate cell division, but only in multinuclear cells.

7. SPECULATIVE CONSIDERATIONS ON THE NATURE OF THE INDUCERS, REPRESSORS, AND ACTIVATORS

While there is abundant experimental evidence that inducers for enzyme synthesis can be normal metabolites, hormones, or their analogs, there is no information as to how these compounds act on the genetic level [11–13, 15, 16, 18–23]. Activation and inhibition of enzymes by substrates or hormones is a well observed phenomenon, but detailed mechanisms of such processes are not unknown [14, 17]. An essential step required seems to be a complex formation between an enzyme and an activator molecule. Here, the structural complementarity between the enzyme and substrate could form the basis for selective interaction. How such selectivity occurs between genes and inducers is not at all clear. There is *in vitro* experimental evidence for complex formation between DNA and various chemical compounds [9, 10]. However, genes in chromosomes are complemented with proteins. Any experimental studies discarding complementary proteins are of limited significance, and information gained from such experiments cannot be applied directly for the interpretation of growth processes. Obviously, the experimental approach in this area is extremely difficult, since *in vivo* studies are not well suited for the exploration of detailed mechanisms of molecular biology.

It is well known that many growth-regulatory metabolites, hormones, and their analogs are small molecules [24–27]. For example, ethylene, the hormone which initiates fruit ripening, is a volatile compound. Also, the growth-stimulating properties of indole-3-acetic acid in plants are well known. The mechanisms of growth activation are obscure. There is good reason to believe that some activation may occur, not at the genetic level, but via secondary structural elements and the ionic trasport system. However, when gene activation occurs, the question is: What are the basic steps and interactions which occur during activation processes ? The specificity mechanism operating between a large DNA structure and a small growth-controlling molecule is hard to visualize. One should consider here that proteins bound to DNA have a wider spectrum of complementary characteristics and could play a regulatory role by multivalent complementation with small molecules [15].

Proteins could be considered to be at least bivalent, being able to interact simultaneously with DNA and also to form complexes by regulatory compounds. The binding of a substrate to a protein could also alter binding characteristics between the DNA–protein complex. Due to the binding, a configurational strain could be exerted to a limited region of the DNA molecule. If this DNA region is functionally involved in RNA synthesis, a local configuration strain could increase or decrease the rate of RNA synthesis. The effect of histones on synthesis has been demonstrated, but the mode of action is still unknown [28-31]. One could speculate that histones with highly polar groups could control large sections of chromsomes, and so activate several genetic units simultaneously. The degree of gene control could depend on the type of histone. Proteins could be more specific and control individual genes. It is evident that extensive studies are required to elucidate the detailed mechanisms of regulatory processes.

8. CONSIDERATIONS ON ALTERATION OF GROWTH PATTERN AND ABNORMAL GROWTH

a. Introduction

The alteration of the cellular growth rate and its characteristics is a phenomenon of particular importance. How and why cells belonging to a particular tissue or organ acquire independence from adjacent cells and establish a new set of growth characteristics is an intriguing problem. The loss of coupling "between adjacent cells" means that the cellular system becomes disorganized, and randomly growing cells lose the functional properties of the original system. Furthermore, altered cells may, in addition, affect adjacent cells so that these also lose their normal characteristics or are destroyed. A progressive, disorderly growth will develop as a consequence.

The cancer problem can be analyzed from this point of view. While the model-system (Fig. 140) is elementary, it nevertheless permits us to focus on the analysis of the abnormal growth problem in a more definite manner than that of generalized verbal argumentation without the aid of schematics.

The fact that cancer can develop under the influence of a variety of physical, chemical, and biological agents is, of course, a very baffling phenomenon. It appears that very basic cellular growth and regulatory processes are involved, and a model-system facilitates exploration of these. At the current level of information, the analysis can be carried out only in a very crude manner.

Basically, one can consider that cellular growth rate is determined by the gross balance of growth-stimulatory and growth-inhibitory entities in the cell. While growth regulation manifests itself at various levels of metabolic processes, the principal determinant of growth control resides in the genetic domain. Activation, induction, and derepression are the principal mechanisms which augment the functional capacity of the synthetic system, and deactivation, repression, and deinduction suppress basically integrated growth. In such processes, many kinds of metabolic entities, such as substrates, hormones, histones, proteins, etc., can participate. However, there is abundant evidence that only a single key substance, when introduced into the cellular system, will increase or decrease the growth rate in some particular tissue [18–22, 31, 32]. Consequently, we are forced to conclude that the complex pattern of regulatory processes is itself controlled by a few key entities through trigger type operations. It further appears that cell growth regulation is under a positive-enforcement type of control. An increase of growth rate seems to require the presence of an activating entity, and growth repression requires the presence of an inhibitory entity. Thus, cellular systems seem not to be under a passive type of control, where a lack of stimulant would represent a condition of growth repression. The evidence of positive growth enforcement control is clearly demonstrated by liver regeneration in conditions of parabiosis. Furthermore, growth control can be exercised at different levels of functional entities. One expects that growth control via the nutrient pool is different from a specific hormonal tissue activation. It appears that the balance ratio between inhibitory and stimulatory compounds would determine whether there is a repression or activation of growth. A passive control mechanism is basically unreliable, nor does it have the proper direct operational characteristics. It appears that a normal cellular system can be controlled in a positive–negative direction, and the level of synthesis is maintained at a level which is determined by the general regulatory processes of the organism. When a cell cannot be regulated by established procedures, an abnormal situation may develop.

b. General Considerations

We shall attempt to speculate, on the basis of the model-system, as to how various agents could produce certain general alterations in cell characteristics via different modes of cellular interaction. One can consider that an internal shift of metabolic balance could be obtained by the blocking and unblocking of various genes. This is a process clearly indicated in the differentiation of cells. Such a condition could arise when external agents enter the cell, where they could interact with genes directly or interact with entities which control gene activity. Intracellular distribution and concentration of metabolites, enzymes, templates, activators, etc., depends not only on the rate of synthesis, but also on how these entities are confined into a limited volume by specific structural elements. Consequently, one expects that the properties of various membranes have a strong controlling effect on basic intracellular processes. This argument applies not only to the cellular membrane, which determines cell volume and boundary, but also to the nuclear membrane, which separates the cytoplasmic and nuclear media. Furthermore, any particular type of functional element confined or integrated into a membrane structure is also a subject of the controlling effect of the membrane characteristics. Consequently, membrane properties which give partial autonomy for a tissue cell or for some functional element appear to have a dominating role in cell growth regulation. There is direct experimental evidence that electrical resistance of the nuclear membrane, which is a function of ionic mobility in the intracellular medium, is high in dormant cells and relatively low in actively growing cells [37]. Furthermore, the effects of the hormone on membrane permeability are well known in the induction and growth processes [38–41]. Whether such hormonal interaction is direct or indirect is not yet clear. However, in many cases a membrane seems to exhibit a basic role in regulating growth.

It appears that intracellular processes depend largely on the functional activity of various membranes which limit the interactions between functional units, and thus control the integrated synthesis leading to growth and division of the cell. However, the external cell membrane, which separates intracellular and extracellular media, not only has an effect upon the cell which it envelops, but can also affect the adjacent cells. While the cell membrane in normal conditions basically controls the flow of nutrients which enter the cell and the metabolites which are excreted by the cell, it also has an important role in the exchange of various molecules and colloidal particles with adjacent cells. These processes are of special significance during differentiation of structures and tissues.

One can consider that cellular interactions may amplify some initial events occurring on the single-cell level. For example, when some particular cell goes through an intracellular metabolic alteration which finally manifests itself within the transfer properties of the external membrane, there can be an enlargment of transport properties. A "leaky" cell can release into the environment, besides metabolites, enzymes, repressors, messengers, templates, etc. These entities may enter adjacent cells and thus influence metabolic and synthetic activity of these cells.

One can formulate the following problem: Assume that a cell has been subjected to the influence of a certain external nongenetic agent. As a consequence, it suffers some metabolic and structural alterations. When the external agent is removed or its activity terminated, how is it possible that the cell retains some of the altered properties and does not regain the initial characteristics entirely? Temporary growth stimulatory and repressive characteristics of hormones are well known, and these effects terminate shortly when addition of the hormone is discontinued. However, a long-term application may have more lasting effects. The basic issue is how to explain the change of cellular properties on a long-term basis without utilizing the concept of genetic mutation. It appears that cellular differentiation processes offer certain guidelines for reasoning. A great variety of functional cells arise from the same embryonic cell by successive transformations. These cells form stable, functional systems, but only on the provision that stable conditions and organization are maintained in the cellular assembly. If, however, a cell is removed from the system and transferred into an isolated medium, the functionality of the cell may be lost. When cells of thyroid tissue are dissociated and are kept in such a state over a certain length of time, these cells lose their ability to produce hormones. Hilfer has shown that when the cells are aggregated before the critical time-limit, they retain their functional property [42]. It appears that the hormone-producing property can be maintained permanently only in an assembly of cells via mutual interaction within a range of critical volume.

One has to consider that in a highly differentiated cell, an effective and stable genetic regulation is possible only when there is a stable intracellular metabolic concentration pattern. This can be maintained in the individual cells only in conditions where the aggregate contains a minimal number of cells. It appears that there is a critical relationship between cell aggregate volume and aggregate boundary. These factors affect the concentration gradients within the cell group. This argument applies to the cells with normal cell membrane characteristics. Suppose that a cell which, for some reason, has altered membrane characteristics is situated in an enviroment of normal cells. It seems that in these conditions the symmetry of intercellular interaction is lost. The cell which has a "leaky" membrane may, in addition, have another dyssymmetry, namely, the exit transport properties may be changed more extensively than those of entrance. As a consequence, there is a loss of functional entities via the cell membrane. If such a loss is extensive, cellular functions may become excessively disorganized and the cell may become nonfunctional. However, at a lesser degree of injury, the cell may acquire a new balance in the distribution of functional entities and establish itself as a new cellular entity. Genetic unblocking may follow, and the cell may dedifferentiate. It seems that cellular transformation is only possible when the cell is able to grow and divide throughout several generations. One may consider that dedifferentiation is not necessarily entirely reversible, since the initial conditions for the basic process are not similar. Differentiation is an orderly and discrete functional reorganization and growth. In contrast, loss of differentiated characteristics via disorderly membrane modification phenomena could produce alterations in cellular characteristics which are in some degree arbitrary. Consequently, many "hybrid" cells could be produced which possess partially predifferentiated characteristics, and partially new properties. Such cells in conditions of further growth could alter and go through further transformations, since they lack the stability of an organized system and stable environment. Furthermore, growth control, which depends on an organized balance of functional elements, may be lost. Disorganized cellular growth may exhibit multiple patterns. In an extreme case, the cells may not be able to grow and will cease to survive. However, in a case where growth regulation is lost, the cell may grow as rapidly as an isolated cell. Here, the limiting factors would be environmental conditions. It appears that cells which arise via such processes may exhibit certain characteristics which depend on the tissues from which they originate. Rapidly growing tissues and organs, such as mammary gland and liver, altered cells could be produced in a relatively short interval, while in connective tissue such processes may require long durations. The potentiality of producing an aggregate of abnormally growing cells would, therefore, be associated with organs and tissues which have a capacity for rapid growth. Removal of growth repression can also be associated with broadening of the metabolic pattern by making use of the large genetic growth potentiality inherent in the system. By genetic and other unblocking processes, the synthetic capability could be largely diversified and enlarged. For example, increased carbohydrate metabolism could lead to increased protein and nucleic acid synthesis. The result would be an increase in the rate of mitosis. A rapid growth rate may lead to the desynchronization of genetic and cytoplasmic division processes, resulting in polyploidy.

Inductive effects in an unbalanced system may lead to progressive changes in the processes of rapid cellular growth. For example, moderate permeability changes in the cell membrane may be augmented during successive cell divisions, leading possibly to several associated phenomena. First, the facilitated entry of substrates and proteins from the cellular environment could activate part of the inactive genetic system. As a result, new proteins could be formed in the cellular system, and the cell could acquire certain new surface characteristics. Secondly, some proteolytic enzymes could diffuse into intracellular regions. These enzymes may operate on external interphases and cause the breakdown of structural elements between adjacent cells. In addition, a rapidly growing "mis-transformed" cell may contain various inducers or derepressors which, in turn, may enter adjacent cells and induce changes by genetic unblocking; so "mis-transformation" could propagate through tissue regions. Furthermore, the change of cellular surface and antigenetic properties will reduce the cohesive interaction between adjacent cells. Consequently, transformed cells may become migratory.

The preceding speculative projection of cellular growth and its abnormalities dealt with the problem in general terms and did not specify how the initial cellular change was triggered. In the following section, we shall attempt to analyze on the model-system the potential role of various entities in the development of abnormal growth.

c. The Effect of Hormones and Specific Substrates

We shall limit our considerations strictly to the model-system and simulate the mechanisms of substrate and hormone actions in very broad terms. The principal modes of hormone action are: a) activation of templates, enzymes, and genetic units, and, b) permeability changes in various membranes and modifications in ionic transport processes. It is well known that various hormones, when introduced into the biological system, will activate various synthetic processes in specific tissues leading to cellular growth and activity. However, while the initial steps seem to be RNA or enzyme synthesis, it is probable that various trigger mechanisms may be operative first [18–23, 38, 43–45]. The hormone type of induction of synthesis reveals a broad action spectrum, not limited to a single enzyme, such as is the case with single-step substrate type of induction. Sequential induction, which leads to multiple enzyme synthesis, represents a secondary phenomenon. It is observed with both the substrate and the hormone type of induction [46, 47, 43]. In order to initiate synthetic processes in a dormant cell, various functional entities have to be activated. One has to consider that a minimum level of internal pools be available for the synthesis, and that some enzymes be present either in active or inactive form.

An initiation process of enzyme synthesis on the model-system in the condition where inactive templates for enzyme synthesis are present requires that a hormone or substrate activate templates (steps 4 and 8, Table IV). Furthermore, it is essential for enzyme synthesis that transport RNA C and amino acid activating enzyme E_t be available. In a dormant cell, amino acid pool P_3 is at a low level. Consequently, transport RNA C is complexed reversibly with polymerase E_1, forming $[CE_1]$ and $[CE_1]^*$. It is sufficient for the initiation of enzyme synthesis that an initial increase of P_3 takes place (steps 5 and 9, Table IV). Growth activation by amino acids appears to be the most essential point, on the basis of model-system considerations. Such a process could be a part of the trigger mechanism. After the initiation of synthesis, a broadening and intensification of the synthetic processes must occur to establish a full-scale growth. This requires activation of several genes for the production of various types of RNA. Polymerases have to be either synthesized or activated from the nonactive state. In order to initiate all these processes, the action spectrum of the inducer must be wide, thereby activating several gene groups (Fig. 140, G_B^s, G_C^s, G_1^s, and G_1^t). This could be accomplished by the direct or indirect action of a hormone with histones and proteins which are associated with DNA. In the model-system hormonal suppression of repressor R_0 formation is indicated for the initiation of chromosomal and cytoplasmic division processes. An integral part of growth activation also involves the enlargement of transport characteristics of nuclear and other membranes. In many cases, this may be the primary step of growth initiation, especially with the synthesis involved in differentiation processes. It appears that a complex pattern of sequential and parallel events occurs in establishing orderly growth.

What could be the mechanism by which hormones or substrates could induce long-lasting changes in functional and structural characteristics in cells? First of all, we may state that such compounds are not mutagenic agents, and genetic mutations are ruled out as a possible cause of cellular alterations. Here again, the basic thesis is that, besides genetic activation, intracellular distribution and concentration of functional entities depends upon the properties of the cell membranes. Transport mechanisms, pinocytosis, and diffusion characteristics of the membrane(s) seem to exert a dominating effect on the cell growth process. While the effect of membrane properties of cell growth during the single division cycle can be analyzed in a straightforward manner, the long-range effect is not so obvious. In order to maintain an abnormal intracellular distribution of functional entities within the cell, drastic changes have occurred in membrane(s) characteristics as well as in the gene activation pattern. These changes have to be maintained after the removal of the hormone or specific growth-controlling substrate. While the hormone is present, cell growth has to continue, and subsequent cellular divisions have to occur in conditions of the abnormal distribution of intracellular metabolities and other functional entities. Synthesis in such a condition could progressively alter the intra- and intercellular distribution of inducers, repressors, derepressors, activators, etc., thus leading to further blocking and unblocking of various sets of genes. Growth in such a condition could lead to cellular transformation in subsequent generations.

How is it possible to maintain altered cell characteristics when the agent is removed from the system? It seems that the cell may regress to the initial state, unless stabilizing conditions are present. Perhaps such a condition can be established when a number of cell divisions has occurred in the presence of the hormone, so that a mass of altered cells is large enough to maintain, by mutual interaction, an environment of altered cells. The size of the cell group has to be large enough so that the volume–surface ratio is favorable for minimizing diffusion losses. Such an autonomous group of rapidly growing cells can lead to further expansion and to interaction with adjacent normal cells. At this point, abnormal growth may become relatively stable, and the course of events can be stopped or modified only by interfering specifically with such a process.

d. External Chemical Agents

There is a wide variety of known carcinogenic agents having different degrees of potentiality for the formation of abnormal growth. No attempt will be made to review the entire subject; we shall consider here only a model-system analysis. One can discard here mutagenic considerations, since it is self-evident that such interactions can, under suitable conditions, lead to a permanently altered cell. Many carcinogens are effective after a single application or through feeding [33, 36]. It appears that a basic requirement for chemical carcinogens is the property of being able to produce complexes with certain functional entities of the cell. For lasting effects, these complexes have to be stabilized and should lead to more permanent bonding. This seems to be especially important when a single application of a carcinogen can have a long-term effect. Consequently, it seems that for abnormal growth development: a) the chemical agent should be bound to a functional element of the cell; b) this interaction has to lead directly or indirectly to an alteration of the growth pattern; c) the effect of the chemical agents has to be maintained for several generations of cell divisions; d) the cell group has to acquire a critical mass before the cells acquire stable and altered characteristics. This seems to suggest that cellular transformations should be stabilized before perhaps half a dozen cellular divisions have occurred. Very little information is available concerning the specific site of action of carcinogenic chemical compounds in *in vivo* systems. However, experiments show that some carcinogenic dyes interact with intracellular proteins [36, 48, 49].

There are various possibilities and modes of cell growth alteration on the nongenetic level. If one considers that proteins and histones, which are part of complementary structures of the chromosome, are synthesized at a different location than is DNA itself, then any molecule which interacts with these complementary proteins before a DNA–protein-complexing process has occurred will interfere with gene function. Furthermore, when a compound interacts with a functional RNA molecule, such as a messenger, then the formation of certain specific proteins may be limited.

The following events could be considered to occur after carcinogen contact and penetration into the cell: 1) direct action of carcinogen with the cell membrane and alteration of permeability characteristics (this is followed by the secondary effects described previously); 2) interaction of the carcinogen with specific protein and protein-forming systems, and interaction with enzymes, repressors, activators, and template-bound proteins. One has to consider that carcinogen-protein and carcinogen–RNA complexes are functionally nonactive. Interaction of chemical agents with many enzymes leads probably to a reduced state of metabolic activity and synthesis. This can be compensated to some degree by cellular regulatory mechanisms; finally, only minor modifications may occur in synthetic processes. Obviously, at large concentrations these effects can be drastic, but it is also evident that large concentrations of interfering molecules lead to cell death. Cell growth modification can be considered only in conditions where cells remain viable. Entities which regulate gene functions are most critical. Consequently, proteins or histones which operate at the genetic level when modified by carcinogens and thereby lose their functional properties can exert far-reaching effects on the cellular growth pattern.   For example, the lack of a repressor means activation of a nonfunctional gene, which, in turn, changes the metabolic pattern.   Inactivation of certain enzymes which participate in regulatory processes may also lead to an unaltered state of regulation.   It appears that only under suitable conditions can the disorganizing effects of a carcinogen lead to successful cell modification. Again, the duration of interference by carcinogens has to be long enough for synthetic processes to be permanently modified. Permanent alteration in cell permeability characteristics can be produced by modifying the action of functional genes which participate in cell-wall synthesis. Such a process will augment cellular transformation characteristics.

e. Viruses

There is a principal difference between viral induction of cancer and induction by various chemical agents. Chemical agents act as modifiers of the existing functional system, while viral agents may, in addition, introduce into the system some new functional elements. These new elements may be either RNA or DNA in nature, depending on the type of viruses which invade the cell.

In order to analyze the viral induction of cancer, a detailed analysis of the intracellular growth processes of viruses in relation to normal cellular growth should be carried out. Unfortunately, this cannot yet be done at the present level of information [50–53, 55]. We shall consider here briefly only the general modes of virus interaction on the model-system basis.   In order for the virus to be a carcinogenic agent, it should not destroy the cell during multiplication, but only modify its functional and structural elements.   It seems essential that some fractions of viral elements be incorporated into the cellular system and maintained there during successive generations of cellular growth. The DNA type of virus, viral genome, could be incorporated into a cellular genetic system. It is not clear whether the attachment occurs directly to the chromosomes, or whether viral units act as extrachromosomal genes. However, if viral genome multiplies during cellular multiplication it is consequently maintained in the system permanently.

We analyze the problem only from a speculative point of view. In order for viral genome to divide, it is essential that the system contain a gene for a specific DNA polymerase synthesis. One can imagine that in the model-system an increase in cellular growth rate could occur when repressor R_0 formation is reduced. A product of viral genome which represses gene G_0^R activity could accomplish this. Furthermore, the activation of genes which control the internal pools seems to be an essential step for increased growth. A key step perhaps is activation of amino acid pool P_3. This occurs when gene G_S^4 is activated, and this, in turn, would lead to activation of the whole synthetic system. What are the changes which lead to cellular transformation? One can consider that permanent changes in cell permeabilities can occur when products of viral genome G_4^V interfere with the activity of functional genes G_1^P involved in the synthesis of structural elements of the membrane. Disorganized membrane structures, in turn, lead to unbalanced regulatory states, and, as a result of this, further blocking or unblocking of some cellular genes could occur. A progressive production of alteration of the genetic activity pattern may lead to the production of several differentiated or dedifferentiated cell types. One could consider that cells which have acquired new stabilized membrane characteristics also acquire permanent genetic activity patterns. The loss of effective regulatory controls could initiate cell types which may lead to cancer formation.

The mode of RNA—virus action on the cellular system in inducing cellular transformation can be considered to be different. How viral RNA multiplication inside the cell leads to permanent modification of cell characteristics is not clear. We can only speculate that, since cellular differentiation occurs in growth conditions where messengers, templates, enzymes, and substrates can be transferred between adjacent cells, one can consider also that the mode of action of viral RNA can be based on mechanisms which are similar in nature. Products of the synthesis of viral RNA which are released have certain functional properties, and these could be incorporated into the functional system of the cell. This may result in the unblocking of genetic activity, which in turn may lead to loss of balance of regulatory mechanisms and alteration of growth characteristics. It appears that a suitable set of conditions is required to establish permanent cellular characteristics. Therefore, we can conclude that while RNA is not a genetic unit, its products can act as blocking and unblocking agents for gene activity and so alter the functional state of the cell.

f. Growth Modification by Disorder

Another way in which the functional state of genes can be modified is the production of breaks in genetic structures. It is well known that viruses, chemical agents, radiation, etc., do produce breaks in the chromosomal structure. The question is, how these breaks could affect cellular characteristics. One can assume that breaks can occur, to a certain degree, in a random manner. If one considers that the interaction between regulatory and functional genes takes place via gene products, such as repressors and inducers, etc., then the local concentration of these products determines the state of gene activity. The local concentration of the repressor in the proximity of the functional gene depends on the distance of the repressor gene from the functional gene. If a break occurs between these two genes, for example, in the model-system between genes G_0^R and G_7^R (Fig. 140), and these genes are moving apart, and the repressive effect of the repressor gene is highly reduced or completely lost, the continuity of the system is lost, and a sequence of gene derepressions may result. Multiple breaks can augment this effect and lead to extensive changes in the cellular metabolic pattern. Of course, only a limited set of random breaks may yield altered cells. The majority of breaks may lead to a complete cellular disorganization and the loss of viability. Chromosome breaks can "heal" during the subsequent synthetic process, but extensive changes may have taken place before "healing" has taken place, and the cell can maintain its modified pattern. This, again, is basically similar to the dedifferentiation process. In general, it appears that chromosomal functioning depends on the geometry of the system, and an agent which modifies the continuity and configuration of genetic elements will affect cellular characteristics.

g. Cell Transformation *in vitro* After Repeated Subculturing

It is well known that, after repeated subculturing for prolonged periods, cells alter their growth characteristics. They may start to grow more rapidly, and finally distinctly altered cells may appear [51, 52, 54]. When injected under proper conditions into a suitable host, these cells may produce tumors. This is indeed an extremely peculiar and baffling phenomenon. Transformation of cellular characteristics via the external medium without subjecting the cells to the influence of a specific interfering agent very strongly supports the conclusion that mutation is not involved in the tumorization process. In the chemical and viral induction of cellular transformation, it was possible to speculate at certain measures on the mechanisms of the processes because one could focus one's attention on definite functional elements and derive certain conclusions from various types of interactions. In contrast, prolonged subculturing of cells in the nutrient medium does not yield any obvious clues for the mechanism of cell transformation and the potentiality of tumor formation. One can consider that a viral type of cell transformation here as a possible explanation of this phenomenon is not substantiated. One can discuss this problem only with great hesitation and uncertainty. A thorough analysis and review has been presented by Eagle on metabolic control and cellular interaction in tissue cultures [52]. Indications are that, at present, it is not possible to unravel the mechanisms which underline the processes of gradual cell transformation which lead to tumorization. Consequently, we can only speculate on the subject.

First, one could consider that cells may possess, after the differentiation process, only a limited period of stability during their lifetime, after which they regress automatically toward the direction of the original state. The process could be speeded up in an artificial medium where proper cellular interaction and orientation is absent. Since there is a random distribution of cell properties, simply on the basis of probability some cells may revert sooner than others and dominate, subsequently, the population of cells. While cellular characteristics arise from the activity pattern of functional and regulatory genes, there is no reason to assume *a priori* that such a distribution cannot be maintained for long durations. However, there is evidence to support the view that in tissue cultures cell stability may be limited only to a certain number of generations and that cells will die unless they become transformed [52]. One is forced to conclude that cell line stability, outside the normal tissues and organs, may be influenced by the composition of the culture medium, or that a differentiated cell dilutes out an important factor during subsequent subdivision. In normal growth in organs and tissues there is a wide range of interactions. Besides absorbing nutrients from the circulatory system, the cell is also exposed to secretions of the whole complex of endocrine secretions and some products of other cells. When cells are transferred to the artificial medium in a random spatial organization in improper contact, the internal balance of the metabolic pattern is likely to be altered. This could automatically cause compensatory or regulatory mechanisms to operate, i.e., this means loss of activation of enzymes, formation of new ones, change of repressors, derepressors, etc. There could also be a gradual shift of metabolic pathways associated with blocking and unblocking of various genes. One can further assume that certain deficiencies in the medium may induce a wider spectrum of functional genes to become operative. It means that the potentiality of the genetic system would be more fully exploited during prolonged growth in the artificial medium than that of the normal environment of the tissues. This uncontrolled enlargement of genetic function may involve activation of certain pathways which have been suppressed during evolutionary development. As a result, new cell characteristics may appear which increase its ability to survive as a single cell. However, this process could be associated with the loss of certain specific properties acquired during differentiation. The process resembles, to a limited degree, dedifferentiation, but essentially it would be a reversion process adapted to individual cell survival. A complementary part of such development would be a change in cell membrane properties. The principal feature of such a cell would be a single-cell economy with minimal restrictions. The basic cellular ability to grow and multiply could be fully utilized. This would lead to an increased flow of nutrients via the external membrane and to a more rapid internal metabolism. An increase in the rate of synthetic processes may require an alteration in the complementary metabolic process, i.e., there may be a change in the carbohydrate or nitrogen metabolism pattern. Consequently, an altered cell would be geared to individual survival and would lose the cooperative characteristics which were acquired during differentiation. When such a cell is reintroduced into the organism from which it originated, its diversification of growth pattern would enable these cells to grow rapidly and independently. Its wide genetic activity spectrum would enable fast adaptation to the new environment. As a consequence, these cells may start to dominate certain regions of the tissues. The character of growth of these altered cells, of course, may depend on the specific tissues or structures at which they originated.

The question is whether these cells can be transformed back into the original or an altered type of subculturing under different conditions. On the basis of the thesis developed here, it should be possible to reverse cellular transformation, but more knowledge would be required about the mechanisms. Since alteration of cell characteristics seems to depend on the distribution pattern of active genes, it seems that a redistribution of genetic activity requires processes which are guided in certain sequences.   Understanding of detailed mechanisms operating during differentiation would be essential for further development in this field.   One can assume that some cell lines may yield extremely stable and other relatively stable cell types.   Basically, no cell can be considered to be completely stable.   One can assume that normal cells can be converted to cancer cells and that cancer cells can revert to less aggressive growth patterns.   However, in order to manipulate such an operation, more detailed knowledge is required about the basic mechanisms of growth processes.

The speculative analysis has consisted primarily of intuitive projection.   The purpose of this discussion was primarily to emphasize the complexity of basic phenomena and to induce speculative approaches in the experimental field.*

9. DIFFERENTIATION

There is little detailed information with regard to specific organizational events initiating differentiation and the subsequent synthetic processes establishing new functional characteristics in the cell.   However, the processes of differentiation play such an important role in growth and development that it seems desirable to consider some aspects of the problem at least in a general way.   Only a crude, descriptive analysis can be carried out.   This involves mostly speculation and raising of issues about the relationship between the growth and differentiation processes.   In order to facilitate the considerations, a diagrammatic presentation in Fig. 141 will serve as a general guide.   Various interconnecting pathways in Fig. 141 are arbitrary and only suggest interaction processes.   One of the

---

* It is of interest to note that Part II of the book was written before Part III was planned and before the literature was reviewed for that purpose. As it appeared later, some of the conclusions reached or proposals made in Section II had been realized in the experimental field already.   This is especially true for certain aspects of plant tumor development and its conversion processes. This, in turn, demonstrates that the speculative approach, at least in principle, may help to orientate one in the selection of proper approaches to the experimental field.

basic problems seems to be the functional interrelationship between chromosomal segments containing genes for growth and differentiation. Superimposed on this sectional interrelationship is the overall regulatory and functional control exerted by genes which participate in the production of histones and regulatory proteins. In addition, the role of cytoplasmic genes should be considered, but there is indeed very little information here, even for speculation.

It appears that there is a difference in the character of the genetic block (repression) in genes involved in the growth process and genes causing the change in cell characteristics. The synthetic processes required for growth could be controlled directly by modifying the activity of growth genes, or regulation could occur on various levels of synthetic units.

In contrast, the genetic events leading to differentiation seem to be distinct and discrete. While the differentiation process in embryonic cells or in some other cells in the unstable developmental stage can be easily influenced by a variety of environmental factors, well stabilized differentiated cells cannot be altered via nutritional channels which affect the growth rate. Maintenance of the differentiated state may require specific environmental conditions, and alteration of the operational characteristic is achieved only on a long-range basis. It appears that the stability and perhaps the mode of blocking is different in differentiated genes than in growth genes. One can consider speculatively that growth genes contain, in a functional sense, a single-step block typical for the induction and repression processes. There is a rather direct functional relationship between gene activity and activator concentration. On the other hand, one may consider that genetic blocks in differentiated cells are controlled in a more stable way. They do not respond directly to normal nutritional changes as do growth genes. Their blocking can be altered only by specific agents or by drastic and prolonged alterations of environmental conditions. This suggests that differentiating genes may have double or multiple blocks, and that these do not respond to normal environmental changes on the basis of

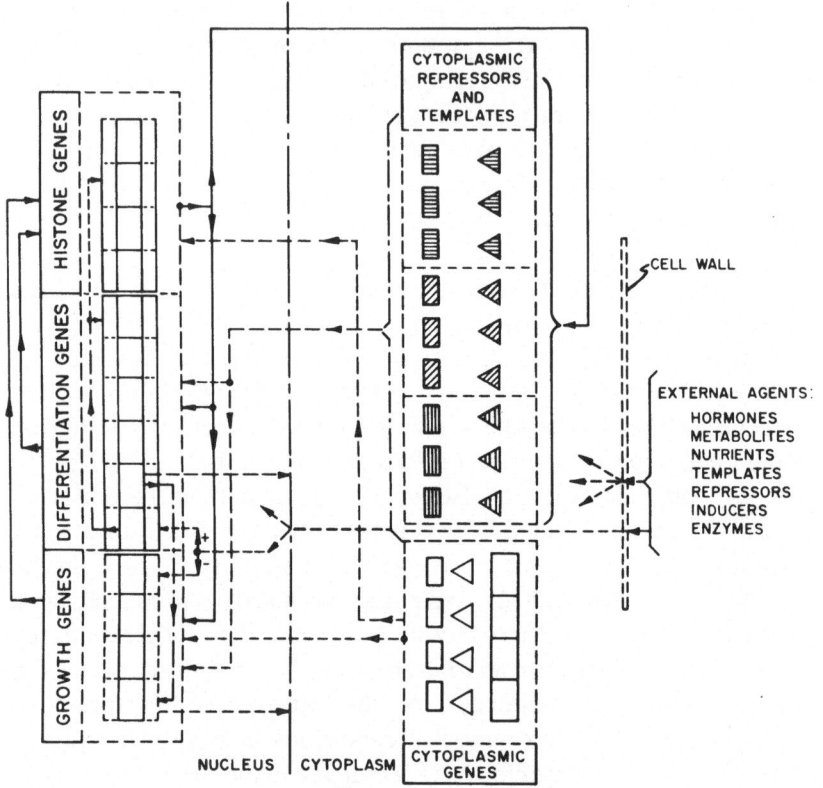

Fig. 141. A schematic outline for the analysis of cellular differentiation.

direct interactions. They seem rather to possess a trigger mechanism which cannot be operated by nonspecific stimuli. Such genetic blocks could be produced perhaps by self-contained metabolic circuits which have to be unbalanced before deblocking occurs. However, once the trigger mechanism becomes active, a multiple set of genetic events could take place, simulating a zipper action operating on a limited segment of chromosome. Consequently, if one considers that initiation of genetic unblocking or blocking is associated with the process of metabolic reactions and synthesis, a specific stimulus has to be introduced to initiate the event. For example, the following considerations could be projected:

a. In an embryonic cell, depletion of some component in the cytoplasmic or nuclear pool could precipitate the event.

b. In a layer of differentiating cells, the thickness of the cellular layer may represent an effective diffusion or transfer barrier for a trigger molecule which originates from a distant basal cell layer; the absence of this element may create the conditions for the initiation of the differentiation process (derepression).

c. A new substrate produced during the differentiation synthesis may lead to further alteration of the genetic blocking and metabolic pattern; this could represent a sequential induction.

d. Introduction of external substrates, templates, enzymes, and hormones may lead to differentiation processes; such a mechanism could be operating in conditions of cellular injury and tissue wounding.

e. Stabilization of differentiated structures may occur when the cell mass has a acquired a critical mass size to contain specific metabolic products in a concentration sufficient to maintain self-contained genetic blocking or unblocking.

There is also a question as to whether the processes of differentiation and cell division are mutually exclusive. One could consider that genes controlling initiation of growth and initiation of differentiation are functionally inhibited to be active simultaneously. Such a condition could arise through a geometric arrangement of genes within the chromosome or by a coupling interaction between genetic blockers. One has to consider here that the differentiating process might be dominating the coupling events. Such mutually interacting genetic elements have been presented by Monod and Jacob [12, 13], and computer analysis has been performed by Heinmets [35]. While one can consider that differentiation is represented by distinct and discrete genetic events, it may be difficult to detect, by visual observations, the initial phase of differentiation, since it may take several cell generations to manifest genetic activity changes in overall cell structure and function.

The exact role of various external agents, such as substrates, hormones, templates, repressors, etc., which may originate from adjacent cells or be carried in by circulation from a more distant source, is still unclear. It is well known that all these agents can initiate new synthesis by induction, activation, and derepression mechanisms, but the initial events and sequence of development is unknown. Interaction of extracellular agents with cellular constituents is not at all straightforward, since there are several membrane barriers (cytoplasmic, nuclear, etc.) which have to be overcome. While all synthesis basically originates from the genetic level, the transport properties of membranes also control genetic activity. Since genetic activity, in turn, depends on intranuclear molecular and macromolecular composition, any agent which alters the membrane structure by direct interaction or interferes with the genetic processes leading to the synthesis of membrane elements can affect genetic activity.

REFERENCES

1. D. Mazia, in The Cell (J. Brachet and A. E. Mirsky, eds.), Vol. III (New York: Academic Press, 1961), p. 77.
2. P. L. Kuempel and A. B. Pardee, J. Cell. Comp. Physiol., 62: 15 (1963).
3. Y. Hotta and H. Stern, Proc. NAS, 49: 648 (1963).
4. H. Stern and Y. Hotta, (1963), in Cell Growth and Cell Division, Vol. II (New York: Academic Press 1963), p. 57.
5. E. Zeuthen, ibid., p. 1.
6. L. Goldstein, ibid, p. 129.
7. J. M. Mitchison, ibid, p. 151.
8. O. Maaløe and C. G. Kurland, ibid., p. 93.
9. M. Zalokar, in Control Mechanisms in Cellular Processes (D. Bonner, ed.) (New York: Ronald Press, 1961), p. 87.
10. B. D. Davis and D. S. Feingold, in The Bacteria, Vol. IV (New York: Academic Press, 1962), p. 343.
11. F. Jacob and J. Monod, (1961), J. Mol. Biol., 3: 318 (1961).
12. F. Jacob and J. Monod, ibid., p. 318.
13. J. Monod and F. Jacob, Cold Spring Harbor Symp. Quant. Biol., 26: 389 (1961).
14. W. E. Knox, Brit. J. Exptl. Pathol., 32:462 (1951).
15. J. Monod, J. P. Changeux, and F. Jacob (1963), J. Mol. Biol. 6: 306 (1963).
16. A. B. Pardee, in The Bacteria (New York: Academic Press, 1962).
17. F. T. Kenney, J. Biol. Chem., 237: 3495 (1962).
18. I. S. Edelman, R. Bogoroch, and G. A. Porter, (1963), Proc. NAS, 50: 1169 (1963).
19. J. D. Wilson, Proc. NAS, 50: 93 (1963).
20. G. P. Talwar, Proc. NAS, 50: 226 (1963).
21. W. D. Noteboom and J. Gorski, Proc. NAS, 50: 250 (1963).

22. U. Hiroshi and G. C. Mueller, Proc. NAS, 50: 256 (1963).
23. W. E. Knox, (1963), Trans. N. Y. Acad. Sci., 25: 503 (1963).
24. S. P. Burg and E. A. Burg, Science, 148: 1190 (1965).
25. A. W. Galston, in The Chemistry and Mode of Action of Plant Growth Substances (New York: Academic Press, 1956), p. 219.
26. H. Linser, ibid., p. 141.
27. H. Veldstra, ibid., p. 117.
28. H. Busch and J. W. Steel, Advances in Cancer Research, 8: 42 (1964).
29. H. Busch, J. W. Steel, S. Hnilica, Ch. W. Taylor, and H. Mavioglu, J. Cell. Comp. Physiol., 62: 95 (1963).
30. M. Izawa, V. G. Allfrey, and A. E. Mirskey, Proc. NAS, 49: 544 (1963).
31. R. P. Perry, Proc. NAS, 48: 2179 (1962).
32. F. T. Kenney and F. J. Kull, Proc. NAS, 50: 493 (1963).
33. C. Huggins, L. Grand, and R. Fukunishi, Proc. NAS, 51: 737 (1964).
34. F. Heinmets, J. Theor. Biol., 60 (1964).
35. F. Heinmets, in Electronic Aspects of Biochemistry, (New York: Academic Press 1964), p. 415.
36. V. R. Potter, Cancer Research, 21: 1331 (1961).
37. J. Wiener, D. Spiro, and W. R. Loewenstein, (1963), J. Cell. Biol., 19: 75A (1963).
38. H. F. DeLuca and G. W. Engstrom, Proc. NAS, 47: 1744 (1961).
39. M. Wickson-Ginzburg and A. K. Solomon, J. Gen. Physiol., 46: 1303 (1963).
40. W. R. Loewenstein and Y. Kanno, ibid., 1123
41. V. R. Pickles and J. F. Sutcliffe, Biochem. Biophys. Acta, 17: 244 (1955).
42. S. R. Hilfer, Devel. Biol., 4: 1 (1962).
43. P. Feigelson, M. Feigelson, and O. Greengard, Recent Progress in Hormone Research, XVII (1962), p. 491.
44. J. J. Holland, Proc. NAS, 49, 23 (1963).
45. J. J. Holland, ibid., 50: 436 (1964)
46. J. Mandelstam and G. A. Jacoby, Biochem. J., 94: 569 (1965).
47. R. Y. Stanier, Ann. Rev. Microbiol., 5: 35 (1951).
48. E. C. Miller and J. A. Miller, Cancer Research. 7: 468 (1947).
49. J. A. Miller and E. C. Miller (1953), Cancer Research, 1: 339 (1953).
50. A. F. Howatson, Advances in Cancer Research, Vol. 8 (New York: Academic Press, 1964), p. 1.
51. R. Dulbecco, (1963), Science, 142: 932 (1963).
52. H. Eagle, Science, 148: 42 (1965).
53. J. W. Beard, in Advances in Cancer Research, Vol. 7 (New York: Academic Press, 1963), p. 1.
54. E. Saksela and P. S. Moorhead, Proc. NAS, 50: 390 (1963).
55. F. Jensen, H. Koprowski, and J. A. Ponten, Proc. NAS, 50: 343 (1963).
56. H. C. Pitot, Cancer Research, 23: 1474 (1963).
57. H. C. Pitot, Perspectives in Biology and Medicine, Vol. VIII (1964), p. 50.
58. H. C. Pitot, in Symposium on Regulation of Enzyme Activity and Synthesis in Normal and Neoplastic Liver, (New York: Pergamon Press, 1963), p. 309.

PART III
Analysis of the Cancer Problem

1. INTRODUCTION

In the first part of the book, we developed a functional cell model which enabled us to establish quantitative interrelationships between the basic functional entities. Furthermore, it was shown that the functionality of the system could be disorganized at various levels, but that the system had a marked ability for self-recovery. In the second part of the book, we designed a more advanced model-system, which made possible a descriptive analysis of genetic and cellular division processes. The importance of trigger mechanisms was emphasized in initiating the sequence of the complex pattern of mechanisms which are required for the orderly development of division processes. Against the background of the model-system data, a speculative analysis was carried out on the general possible mechanisms and events which could produce abnormal growth on the cellular level.

In this section, we shall attempt to review briefly some limited, but specific, experimental material in the field of cellular growth, in the areas which may have direct or indirect bearing on cancer development. Our principal aim here is to elucidate some basic phenomena of cellular growth on the basis of varied, and often not interconnected, experimental data, in order to clarify the specific mechanisms which could be utilized for the integration of isolated processes into a unified system. Thus, an attempt could be made to analyze the problem of cancer from an experimental as well as a theoretical point of view, and projections could be made using both types of information simultaneously.

2. COMMENTS ON ABNORMAL GROWTH

Previously, we analyzed normal growth from the point of view of a model-system. Here, we shall review some of the experimental work described in the literature, and then try to formulate opinions on the basis both of this experimental work and of previous analysis on model-systems. While speculative, the primary aim of the present analysis is to try to derive and unify certain underlying features which lead to cancer. This unification requires data from mammalian as well as from plant systems. It is expected that the phenomenon of abnormal growth, which is widespread in nature, contains general underlying features that are present in all viable systems. It is impossible to analyze the growth of cancer on the basis of the enormous amount of data collected in the literature, which is highly heterogeneous, unclassified, and chaotic in character. This fact in itself reflects the difficulty of the subject. We should, however, prefer to explore a few isolated works which lend themselves to a more careful analysis, provided that these studies confine themselves to a fairly long phase of research on a rather well-defined system. This means, for example, that the tissues of a specific cancer should be studied under well-defined experimental conditions, in which the same species is used throughout, and a comparative analysis with the normal system is carried out. Such works are to be found but rarely. We have selected two principal studies performed in the field of mammalian and plant tumors.

3. MAMMALIAN TUMOR

Extensive studies have been carried out on mammalian abnormal growths by Pitot and his associates. Numerous publications, of which only some are cited here, deal with biochemical, physiological, and also theoretical analyses of cancer [1–8, 17]. Their work is mainly concerned with the so-called minimal

deviation hepatoma, which exhibits nearly all of the morphological and biochemical characteristics of normal tissue, and
can thus be considered an abnormal growth with minimal deviation from the normal state. There is no aerobic glycolysis,
and the cell possesses the normal diploid number of chromosomes. These studies reveal that, while many enzymes
operating in a normal cell and a minimal hepatoma cell are
similar, they may perhaps be modified on the quantitative level.
It is difficult to differentiate the normal cell from the cancer
cell. The principal difference seems to be within the regulatory mechanism, since induction of tryptophan pyrrolase seems
to be limited or inhibited in the hepatoma cell. The effect of a
lack of a regulatory mechanism in at least some areas of cell
metabolism has been correlated with altered template stability.
Pitot, on the basis of analysis of regulatory circuits, has
suggested that abnormal growth can be interpreted in terms of
template instability. While penetrating studies of this group on
various aspects of this type of tumorization are highly informa-
tive, nevertheless, the basic issues are unclear. In summary,
I should like to quote Pitot's closing statement: "In the final
analysis we suggest that alterations in template stability at the
molecular level can, at the cellular and tissue levels, lead in
one way or another to the thousands of biochemical, patholog-
ical, and clinical syndromes we associate with the disease,
cancer. First, altered template stability becomes the molecular mask of malignancy." It appears that cancer represents
autonomous growth, but that the degree by which it deviates
from the growth of the host tissue cannot be determined at the
level of a single entity. There are, however, patterns of
variation.

Nutrition of cancer cells is an integral part of this problem.
Therefore, we shall review some other systems where nutrition studies have been carried out. There is evidence that
cancer cells have different nutritional requirements from those
of the normal cells from which they develop. Studies with
single-cell technique have revealed that there is a highly
varied spectrum of nutritional difference between the normal

cell and the cancer cell [9–12]. It appears that the malignant
S3 Hela cell is nutritionally much more self-sufficient than a
normal diploid cell, and, in fact, this characteristic may be
associated with malignancy. Furthermore, growth of the
mammalian cell in a tissue culture reveals also that cells
which have passed through a transformation, after prolonged
subculturing, may also acquire different nutritional characteristics, in the sense that the cancer cell can synthesize some
metabolites and thus escape from the normal control mechanism.

4. PLANT TUMOR

Extensive and penetrating studies on some plant tumors
have been carried out by Braun and associates. The subject
has been thoroughly analyzed in several reviews [13–15], and
the basic conclusions seem to be that the tumor cells are
altered cells which are randomly proliferating, and which
reproduce true to their own type, and against the growth of
which the host organism has no adequate control or even protection. Therefore, the tumor cell, being an altered cell arising
from a normal cell line, is not derived by genetic mutation. A
basically similar view was expressed by Pitot [2, 3]. The
mechanism by which a tumor cell can arise in a tissue containing normal cells is a very difficult one to unravel. Crown-
gall tumor, which is initiated by the bacterium *Agrobacterium
tumefaciens,* does not develop via mutagenic events, but variations and changes in the genetic activity pattern are indicated.
Braun has carried out particularly interesting studies with a
line of tobacco-pith cells. These cells, under the influence of
wounding and bacteria, are transformed into tumor cells in a
short time. Once this transformation has occurred, the abnormal prolification of the tumor becomes an automatic process. These cells are completely independent from the original
inciting agents, and can cause secondary tumors in different
areas of the plant. The process of tumorization involves the
activation of a series of biosynthetic systems, and the rupture

of cells in the wounding area seems to activate the normal cells in that region. These events lead to a refractory transformation phase. When, however, the tumor-inducing principle, which is a component from the bacterial system, is present, the cell will be permanently transformed. The degree of transformation is conditional and depends on the inducer concentration. Lippincott and Lippincott have presented evidence that the tumor-inducing principle is unstable [16]. The authors suggest, on the basis of thermal inactivation data, that the tumor-inducing agent may be nucleoprotein. However, the processes of synthesis and transport may be indicated as well.

One of the most interesting features of the tumor cell is the alteration of nutritional requirements. The tumor cell is capable of growing indefinitely in a limited medium containing inorganic salts and sucrose, while normal cells are not capable of growing there [13–15]. Furthermore, it is evident that the tumor cell is capable of synthesizing certain growth hormones, while normal cells cannot. When cellular transformation is only partial, then supplementation with certain essential metabolites can increase the growth rate in those cells. All this evidence shows that there is a wide pattern and variation in the degree of transformation, and only those cells which have gone through an extreme transformation establish a very independent type of growth which is not easily influenced by other factors. There is evidence, however, that an independently growing cell can be reconverted when grafted back into the plant, and that it can again acquire normal characteristics. The very fact that a cancer cell can be changed under the influence of normal tissue shows definitely that the transformation is not associated with genetic mutation, but may involve genetic activity changes. It is of great importance that the same conclusion is reached by Braun studying the plant and by Pitot studying mammalian tissues. It is also interesting that the concept of the genetic activity pattern as being a cause for cellular transformation could also be derived on the basis of model-system analysis. In summary, it seems permissible to conclude that mutagenic events are not required for the tumorization process.

5. BACKGROUND OF THE CANCER PROBLEM

On the basis of presented experimental evidence and model-system data, we shall try to carry out a general analysis of the abnormal growth process. This will be done in an unsystematic fashion, because there is not enough experimental material to formulate the basic mechanisms and thereby derive a generalized concept of cancer. However, as we have seen in the previous section, there is enough information to show that cancer indeed represents an aberration in the metabolic pattern and synthetic activity as well as having the characteristics of reversibility. This shows that we are dealing with a highly complex set of processes, which could perhaps be speculatively analyzed on the basis of information from the literature, supplemented by our own analysis on model-systems, thus presenting a new approach to the problem. Various theories have been proposed to interpret the cancer mechanisms [2, 3, 17], but no attempt will be made here to review these proposals because they are all rather arbitrary and not well defined.

We shall review the cancer problem on the basis of various bits of information arising from different areas of biology, and shall try to elucidate certain specific mechanisms and to postulate basic concepts for the development of cancer.

The autonomously growing cancer cell can have varying levels of autonomy, as manifested by changes of nutritional requirements, alterations in regulatory mechanisms, and degrees of reversibility. There are numerous inducing agents which can produce abnormal cellular growth characteristics. Before attempting to discuss the mode of cancer induction in detail, we shall review some more basic aspects of cancer, which have bearings on normal growth and development. As has been shown previously, the cancer cell cannot be differentiated from the normal cell on the basis of chromosomal abnormality [2, 3], since diploid cells can have tumor cell characteristics. Also, the cancer cell cannot be identified on the basis of growth rate, because, while many cancers grow fast, they can also be rather slow, not exceeding the rate of many rapidly growing normal cells. Furthermore, the cancer cell does not have to go through successive dedifferentiations or differentiations, which are, of course, complementary characteristics of a complex tumorization process. But this is not the essential requirement, because a tumor cell can still be effective in producing specific functions and secretions [12]. For example, hormone production can still be carried out by the tumor cell. So it appears that there is no distinct difference between the cancer cell and the normal cell, except for some minor deviations, such as in carbohydrate metabolism, but these deviations can be extremely small, and their significance has been brought into question [2, 3]. The question then arises, why is the cell a cancer cell? It appears that the principal symptom of the cancer cell is that it is independent and is not under the influence of the neighboring cells. One can consider that cancer is a social disease of the cellular society. In tissues, normal cells are *all* integrated into a society of cells which are under mutual influence. This mutual interaction is the basic requirement for an organized cell system. This is in contrast to a random cell system, where there is extremely loose coupling between the cells, or where this coupling has been entirely lost. The organization of cellular systems and complex organisms is derived through a very orderly sequential process where formation depends from the very beginning on the interaction of cells which grow and develop and differentiate. This interaction property, then, is, in our opinion, the most important property for organized growth, and the loss of interdependence can be considered to be associated with tumorization of the cell. Before we continue this analysis, we shall review briefly some basic processes which are associated with tissue formation.

The development of tissues and organs is an extremely complex process, and the character of this developmental system makes it difficult to elucidate. The detailed molecular processes and elements which operate such systems have not yet been clearly identified. Consequently, many conclusions are tentative and speculative. Nevertheless, the general impression is that substantial progress has been made in the formulation of certain basic principles. Cellular organization depends on cell movements, selective cellular adhesions, and mutual determinative processes appearing between cellular aggregates [18–20]. Specific cellular interaction phenomena have been elegantly demonstrated in tissue cultures by Moscona [21, 22]. In addition, he has established that surface properties and cellular interactions depend on intercellular synthesis. It appears that the organization of cells into the tissue patterns is controlled by a variety of environmental factors, and that the singular cell expression *per se* is not sufficient to give rise to the wide variety of development. Nevertheless, the cell has to have genetic potentiality to be influenced by its environment. In order to build a complex structure from small entities or units, it is essential that those entities be combined as aggregates, and that a certain type of force operate between the individual entities. The aggregation of specific cell lines indicates that there is a high degree of specificity and complementarity among various cells [21]. Factors which control the specificity are not well understood at present, but there are further indications, as pointed out by Moscona, that certain globulins and lipoproteins may be operative on the surface. Since intercellular adhesiveness is essential for the maintenance of an organized cellular community, it is desirable to review this problem further. The following questions could be asked:

1. What chemical factors in the intercellular space affect adhesiveness and cellular mobility?
2. What types of forces operate between adjacent cell boundaries, and what kinds of linkages are formed between the cells?
3. What are the nutritional factors leading to cell adhesiveness?
4. Are there some inductive effects by molecular species operating in the intercellular space, and do these molecules control the synthesis of the surface structure?

5. What are the specific changes in surface properties during cellular transformation and in the process of tumorization?

The data in the literature are rather heterogeneous and do not provide answers to these questions. Nevertheless, many specific problems have been studied and provide definite information pertaining to certain cellular interactions during growth. It is evident that the direction of cell movement (chemotactic effect) may be influenced by substances present in the medium. Harris has shown that malic acid and some other substrates produce a chemotactic effect in spermatozoa [55]. Malate ions are reversibly absorbed in cellular receptors. It appears that chemotactic compounds are specific for the cell type. It has also been shown that some well-characterized polysaccharides and starch granules exhibit a chemotactic response. These observations are of great importance, and we shall return to this subject when we analyze inductive mechanisms.

In contrast, animal cells move by amoeboid motion, and only when in contact with solid or semisolid surfaces [55]. Since there is little substantial information available about the mechanisms of this amoeboid motion, the effect of chemotactic compounds on such cells is not clear. It is speculated that a chemotactic compound produces pseudopodia at the exposed side of the cell, or that, perhaps, the substrate induces the protoplasm to flow preferentially on the exposed side of the cell and helps the protoplasm to penetrate into the existing pseudopodia. Clarification of such problems is of great importance for understanding the processes of organized growth. Abercombie has performed various studies and has postulated an hypothesis on stimulus–response mechanisms of fibroblast locomotory orientation [23]. He considers that the locomotory organ is a ruffled membrane. The fibroblast moves with changes of direction by the waxing and waning of the ruffled membrane at different regions of its margin. The ruffled membrane can be inhibited by contact inhibition and contact guidance respectively. The normal chick heart fibroblast shows characteristic locomotory behavior *in vitro* but it is dominated by mutual contact inhibition and contact guidance. This seems to be the general fibroblast pattern, but each cell type probably has individual behavioral responses. In contrast the *malignant* fibroblast has either reduced or absent contact inhibition, but exhibits no loss of contact guidance. The lack of inhibition by contact with *normal* fibroblasts permits the sarcoma fibroblast to move away from or toward the normal colony of cells, but it cannot "stick" to the other cells. This means that the sarcoma fibroblast exhibits the property of substrate induced locomotion, but the surface properties of the ruffled membrane have been altered, so that a permanent adhesion cannot be developed.

It is of interest to note that wandering cells, macrophages, and leucocytes behave like sarcoma cells, being immune to contact inhibition by normal fibroblasts. The evidence that cells can "sense" the presence of a cellular substrate and adjust their movement accordingly is well demonstrated in *in vitro* experiments. When fibroblasts with a low density population, moving on a glass substrate, reach a part of the glass surface from which another fibroblast has been removed, they move away. This suggests that a specific amount of the fibroblast-specific substrate had adhered to the glass surface to be detected by an oncoming cell. In a densely populated region, however, the cells are unable to move away, and establish adhesions that may last for hours. In populations, the dispersal occurs in a directional way, from a populated to an unpopulated region, and cell movement shows a bias toward areas which are free of fibroblasts of a specific substrate. On a plane surface, the monolayer tendency is persistent, and the more effective contact inhibition is, the more perfect is the monolayer. It appears that monolayering features in cell cultures reveal properties of cells which are essential for the formation of organized structures, while randomly growing tumor cells have lost this property. The loss of orderly growth in virally induced cancer cells has been demonstrated by Dulbecco [24]. Furthermore, it has been shown by Klein and Klein and by

Purdom et al. that there is a reduction of cellular surface
adhesions in methylchloranthrene-induced sarcoma cells, and
that there is an increase in surface charge [25, 26]. Moreover,
these cells exhibited increased invasiveness and metastasiz-
ability. In addition, it was observed by Klein that new antigens
appeared on the cell surface after the polyoma-virus induction
of cancer [27]. It has been further shown that cell movement
and adhesiveness are reduced by woundings or by disaggrega-
tion, and only slowly return to the normal state as repair or
reaggregation occurs [28]. This time-dependent property can,
perhaps, offer some explanations for the mechanisms of tumor
formation in plants.

The adhesive contacts between cells, which in further
development lead to a more permanent type of bonding, represent, perhaps, a set of synthetic processes, which may be
initiated by intercellular interactions. It is far from clear how
adhesive structural elements develop, and what their fundamental properties are, besides those of providing physical contact between the cells. Fawcett has reviewed and analyzed
these types of cellular interactions [29], including the formation
of intercellular bridges. This term originally described the
connections between epithelial cells, but similar associations
have been observed between some other types of cells. True
intercellular bridges establish protoplasmic continuity between
the cells, and they may have a role in maintaining synchrony of
differentiation within the group of conjoined cells. Grobstein
[30] has focused attention on cell contact and movement in relation to embryonic induction. The mechanism involved in such
processes can have far-reaching significance in interpreting
the events of cellular interaction during growth. While structural contact exists between adjacent cells at some points,
most parts of the cell surface are at a certain distance from
one another. Since some tissues can be dissociated into individual viable cells by treatment with cation-depleting agents
[21], this indicates that forces operating at the bonding sites
are essentially of ionic character. It is also indicated that
inductive substances can act from a certain distance and via

membrane filters [30]. While it has been established that there is no general requirement for direct contact between inductive interactants, it appears that at least some inductive processes operate only in a short-range region. The interspace between adjacent cells is not at all clearly defined. There is evidence that there is an amorphous zone about a few hundred Angstroms wide between the cells. Brandt [31] has presented evidence that the extraneous cell "coat" participates in the transport process (pinocytosis) of molecules from the extracellular region into the cell. Molecules attach to the cell "coat" and are subsequently "processed" into the cell by mechanisms which are not yet clear. Compounds which exhibit a high degree of complexing property dissolve the coat during the process of pinocytosis. This fact can perhaps provide an explanation for the increased cellular permeability resulting when cells are exposed to solutions containing polypeptides and polysaccharides—hormones.

It appears that, in induction processes, one may not need a direct transfer of inducer molecules from one interacting tissue to another, but through the interaction of surface-associated materials, which lead to new properties of the interspace, and hence of the involved cells. Such new properties could arise from the polymerization of interspace molecules or by the complexing of molecules of both sides, leading to macromolecules. The implication here is that the microenvironment may have an initial critical role in the events of induction. The significance of intercellular macromolecular exudates leading to integration mechanisms has been re-emphasized by Moscona [21], and it appears to us that the underlying biosynthetic processes can play an important role in the development of tumorization of cells.

Since experimental evidence reveals that adhesive, orientative, and locomotive properties are of basic importance in the development of integrated cellular systems, the factors which are essential to the formation of external structural elements of the cell should be investigated. It appears that cellular nutrition and the composition of the cellular media have a significant effect on cellular growth characteristics in the tissue culture medium. At the present time, however, the subject is still under study, and available experimental data are partially controversial and confusing. Nevertheless, nutritional studies may provide some principal clues for unraveling the basic mechanisms leading to the adhesive properties of cellular growth, and thus may yield insight into processes of cellular transformation leading to the cancerous state of the cell.

The attachment and spreading of cells in a tissue culture has been reviewed by Taylor, and a series of experiments on the subject has been carried out [32]. It is evident that environmental factors (such as the material of the substratum) and the components of the medium have effects on cellular growth and morphology. It appears to us that studies of attachment and spreading of cells on a glass surface or other artificial surface layers do not provide specific clues on intercellular interaction mechanisms, while experimental data concerning cellular growth in terms of direct cellular interaction are of more importance. The effect of certain serum fractions on cellular growth has been a subject of intensive studies, which have been reviewed by Eagle and Levintow [12, 33, 34]. While the mode of action of some essential serum components is not clear, their effect can be clearly shown in tissue cultures. It is evident [32] that serum retards the spreading of human conjunctiva cells in the substratum and affects the morphology of the cells. These cells are roughened by many small protrusions, which are constantly in movement. It was further observed that serum microexudate of living cells and other macromolecular substances inhibited cellular attachment and spreading.

Cellular adhesions are the primary characteristic of cell association. It is of interest to study this problem in more detail and consider it from various points of view. It is not only a matter of an aggregate formation on the basis of a physical–chemical type of interaction, but there is also a relationship between external properties of the cell and intracellular functional processes. The observation that a glucoprotein fraction in the serum is essential for cell adhesions seems to have great significance [9, 21, 35]. While some studies show that a macromolecule of serum component is not essential, and that a small molecule derived from a macromolecule is sufficient [33, 34], this does not necessarily represent a contradiction. As a matter of fact, it may offer suggestions for the basic operational mechanisms. We should like to explore this problem further. While isolation of the micromolecule has been difficult, the macromolecule has been isolated as a_1-glucoglobulin [9, 35]. It is of interest to note here that this glucoprotein fraction affects also the adhesion of collagen [37]. Experiments indicate that this mucoprotein is involved in the cross-linking of thin elements of acid-soluble fibers. We consider it of particular importance that a compound observed to be active in cellular adhesion is also active in collagen aggregation. While this could be a coincidence, it is worthwhile to explore this problem further. The question can be asked: What is the role of the glucoprotein in the process of cellular adhesions? Is that strictly a problem of physicochemical complex formation and thus a facilitation of the adhesion process, or is there some other functional property operational in the living biological system? The fact that collagen fibers are formed in the presence of mucoprotein under conditions where there is no synthesis, and that some other compounds are also effective, seems to suggest that it acts as a cross-linking agent in this type of system. The fact that it will also affect cellular adhesion can be considered from three distinctly different points of view:

1. It could also help cellular adhesion as a physical — chemical agent.
2. It could affect some facets of cellular synthesis.
3. Both phenomena could operate in adhesion of cellular growth.

The fact that response is relatively rapid [35] suggests that glucoprotein may have a complexing role. However, the evidence that cells in a tissue culture do maintain their adhe-

sive properties only temporarily when some nonspecific adhesive compounds are added suggests that the role of glucoprotein
may be more diverse than is indicated by this simple experiment. Therefore we shall speculate on some other functional
roles of a_1-glucoprotein. Work in bacteriology shows that
polysaccharides act as enzyme inducers [38, 39, 40]. It has been
shown by Hughes and Jeanloz [36, 41] that *Diplococcus pneumoniae*
cultures contain several extracellular enzymes, including neuraminidase and β-galactosidase. The neuraminidase releases
N-acetylneuraminic acid (sialic acid) and the a_1-acid glucoprotein of human plasma. It is evident that glucoprotein serves as
a substrate for this enzyme. After sialic acid has been released
from a_1-acid glucoprotein, the enzyme β-galactosidase releases galactose. Subsequently, this glucoprotein residue
serves as a substrate for the enzyme β-N-acetylglucosamini-
dase, which releases 2-acetamido-2-deoxy-D-glucose. It is
evident that degradation of a_1-acid glucoprotein requires a
series of enzymes and that they have to act in a definite sequence in order to release subunits. It is of great interest to
note that the enzyme neuraminidase, in the treatment of fetuin,
causes complete loss of its biological activity [9]. This is of
particular interest, since both fetuin and its serum contain
factors essential for mammalian cell adhesion and growth. The
observation by Moscona that a_1-globulin is also effective in
promoting the adhesion of embryonic cells, while less effective
than a-liproprotein, is of great importance [21]. Unfortunately,
in various experiments reported in the literature, the work has
not been carried out with highly purified serum components.
Therefore, an exact comparison of various experiments may
not be possible, and the use of this data for the interpretation
of basic processes has to be made with certain reservations.
Many of the serum fractions used have been isolated and identified previously [42, 43, 44], and, therefore, the experiments
seem to have a general application. Since different cells lines
may require different serum fractions, further experiments are
required to elucidate these points. In this respect, the nutritional study of cancer cells would be of particular interest here.

Let us assume that " a_1-glucoprotein" is an essential component in the serum of fetuin for a particular mammalian cell
line for growth in a tissue culture [9], and let us pose the question of what functional role it plays in this process. Does it act
solely as a colloidal agent in implementing cellular adhesions,
or does it have another role? The evidence that removal of
sialic acid by neuraminidase abolishes the functional activity
can be interpreted in several ways. On the current level of
our information, we can analyze this problem only in a highly
speculative manner. Since the removal of a small sialic acid
molecule from a macromolecule causes the loss of its activity,
one can definitely conclude that the general colloidal property
per se is not the only property required for growth-promoting
action. The colloidal and viscous property, however, while
important, may be secondary. Perhaps the interacting properties of a macromolecule for attachment could have been
highly reduced by the removal of highly polar sialic acid. On
the other hand, sialic acid may be required for a highly specific
interaction. Some clarification of this problem may be found in
the studies of influenza virus interaction with the cell membrane. The virus contains the enzyme neuraminidase and
attaches itself via this enzyme to neuraminic acid residue on
the cell membrane structure [45]. Sialic and neuraminic acid
are very closely related structurally, differing only by one
acetyl group. Neuraminidase also liberates sialic acid from
mucoproteins, and later loses its virus inhibitor property.
Furthermore, interaction between inhibitory mucoproteins and
a virus results in the loss of the inhibitory activity of the
mucoproteins.

In view of such experimental data, a few questions could be
raised as to the role of a_1-glucoglobulin leading to cellular
adhesion. Since the removal of sialic acid resulted in the loss
of biological activity of a_1-glucoglobulin, it seems that the
former may act as a primary site of interaction with external
sub-units of the cell membrane. At this point, one can pose a
question: Would it be possible for a macromolecule to act as
a binding element, linking two adjacent cells together? While

it has been determined that sialic acid is an essential molecule in the system, how would it be possible to produce a cellular association between two similar cells? It appears that the simplest assumption that could be made is: the cell surface could have two separate interaction sites or enzymes, one of which would be a sialic acid type, and the other a peptide-specific enzyme, corresponding to the terminal part of the peptide portion of the glucoprotein. While such an assumption does not seem too far-fetched, it is too simple, and some additional experimental data should be explored. It has been shown that requirements for serum proteins can be replaced by polyvinylpyrrolidine or dextrin. For mammalian cells and for mammary carcinoma, which required an umbilical cord extract, a mucopolysaccharide could replace the extract, while polyvinylpyrrolidine could replace it only partially [34]. On the other hand, it has been shown that macromolecules as such are not invariably required, either for growth of mammalian cells or for their attachment on the glass, but that serum protein acts as a supply for one or more essential growth factors of relatively low molecular weight, liberated by proteolytic enzymes. These small molecules, derived, for example, by the action of pancreatic extract from serum, can penetrate a membrane and support cell growth indefinitely [33, 34]. The role of some carbohydrates is indicated, since Hela cell growth was stimulated when a solution containing glucose and phosphate was heated and added to the culture medium at low serum concentration. It should be pointed out here that many cells adapt readily to a limited growth medium, and such cells cannot be utilized to study primary interaction mechanisms for the systems containing normal cells. The role of glucoprotein in the process of cellular growth in the tissue culture is perplexing, and various experimental data seem to be in contradiction. Before final conclusions are drawn, a further role for glucoproteins should be considered.

In view of the fact that glucoproteins serve as substrates for certain degrading enzymes, the question can be asked whether glucoprotein can serve as an enzyme inducer. While there is no definite information on this subject, there is considerable information on the induction of enzymes by polysaccharides. Since a polysaccharide chain is a part of α_1-acid glucoproetin and serves as a primary target for enzyme action, it is considered worthwhile to explore further the phenomenon of enzyme induction. The mechanisms of enzyme induction by polymeric inducers is far from clear. This subject has been reviewed by Pollack [38], and there are several possibilities to consider. One could consider that there is an initial one-step degradation of the substrate by an already-existing enzyme, followed by the induction of a new enzyme by a released monosaccharide or by another small molecule. This process could repeat itself until all necessary enzymes are formed for the degradation of the polymeric substrate. On the other hand, the substrate could be incorporated into the cell by pinocytosis, and the whole process could take place inside the cell. While it is not clear which of the sites of enzyme induction is operative, there is evidence of a sequential appearance of enzymes in bacteria after polymeric induction. Studies by Barker and associates, [40, and personal communication] on the induction of bacterial enzymes with polysaccharide indicate a sequential induction. However, the sequence of appearance of the enzymes is not linear in relation to the alignment of monomeric units in the polymer. Since enzymes which attack the polymer at nonterminal sites appear during sequential induction, it appears that pinocytosis may be operative as a part of the inductive process. The role of glucoproteins is further diversified when one considers that many glucoproteins are hormones [43] and control growth. There is a possibility that the action of hormonal glucoproteins and cell glucoproteins may have rather similar physiological action mechanisms as far as the cell membrane is concerned. The effect of an extracellular polymeric substance on intracellular synthetic mechanisms raises many further questions and is of great interest. The classical type of sequential enzyme induction operates on the principle that a small substrate molecule acts as an inducer for the enzyme. The induced enzyme converts the inducer into a second substrate, which in turn acts as the inducer for the next enzyme [46,47]. Sequential induction may lead to the formation of a number of enzymes. For the polymeric inducer, however, we must consider a more complex type of process. Substrates released from the polymeric inducer after the induction of degradative enzymes can act again as inducers for sequential induction, thus opening up new metabolic pathways. It appears that a polymeric substrate can contain a large program for the organization of complex metabolic processes and pathways.

Whether such an inductive property appears in the role of α_1-acid glucoprotein as the organizer of adhesion processes between the cells remains yet to be proven. However, the evidence that proteolytic products of serum and other small molecules promote cellular adhesions and growth [33,34] suggests this possibility. However, this problem should be studied further. While it is evident so far that the polysaccharide type of induction can lead to a set of new degradative enzymes, perhaps the reaction products also induce some new enzymes, which could be instrumental in leading to cross-linking and a more permanent type of linkage formation between the cells. Furthermore, the role of the polypeptide portion of glucoprotein should be explored. Maybe this portion of the molecule could provide an additional programming for the polymerase type of enzymes. A plasma clotting agent had been isolated from bacterial cultures [38], which suggests that enzymes or their products are produced by cells which carry out the process of polymerization. It has been shown that heparin and related polysaccharides, when present in a bacterial culture medium, give rise to several degradating enzymes [48]. On the other hand, heparin seems to have a many-sided biological role. For instances, it has been proposed that heparin is metabolized by connective tissue cells, and thereby stimulated to produce a specific mucopolysaccharide as a contribution to the formation of the extracellular ground substance, which further leads to connective tissue formation [49,50]. The evidence that various glucoprotein and mucoprotein levels increase in plasma during neoplastic and other chronic diseases [43, 44] suggests a role for these substances which is related to cellular growth and breakdown. It has also been observed that extensive wound healing is accompanied by a rise of the a-globulin content of mammalian serum. Similarly, whole-body radiation in mice produces an increase in a-globulin [9]. The preceding experimental evidence strongly suggests that the polysaccharide moiety of macromolecules may play an essential role in the processes of intercellular adhesion formation. Therefore, it is of further interest to follow the role of protein polysaccharides in the development of intracellular fibrous structures and aggregates. Jackson has reviewed extensively the problem of connective tissue development and its structure [51]. We use this review as source material, but consider here only limited aspects of this subject, being primarily concerned with the role of intracellular fibrous structures and ground substance.

The cells are bathed by an interstitial fluid which contains plasma proteins, electrolytes, hormones, and other circulating materials. There are small, significant amounts of glucoproteins, and in some tissues sialic acid is a diffuse constituent of the ground substance. A variety of acid mucopolysaccharides are present, depending on the type of tissue, species, age, etc. Their functional role is not clear, but it is suggested that mucopolysaccharides play a prominent functional role in the physiology of the cell surface. In the native state, however, most of the mucopolysaccharides are combined with proteins, and, hence, their function should be considered with this fact borne in mind. The synthesis of polysaccharides from monomeric units requires one or more enzymes, and glucose seems to serve as an important source for sugar units. The polysaccharide is firmly linked to the protein, and drastic treatment is required to split such a linkage. Jackson points out that proteinpolysaccharides are readily degraded by several proteolytic enzymes, notably by trypsin, papain, and streptolysin, causing, in solution, a marked reduction of viscosity. Hence it is indicated that this moiety is instrumental in macromolecular aggregation. There is not enough specific information, but it is suggested that the protein moiety has properties similar to globular proteins; here again the amino acid composition depends on the origin of the tissues. Electron microscope examinations reveal that papain treatment of protein-polysaccharides destroys macromolecular order and causes the disappearance of filamentous units.

The mechanisms and steps for the synthesis of protein-polysaccharides are not at all clear, but it is assumed that both moieties are produced within the same cell. Furthermore, it is assumed that some other substances, e.g., sialic acid, may form part of the proteinpolysaccharide macromolecule. How a proteinpolysaccharide is assembled is not clear, but there is evidence that in some cells it is synthesized as a unit within the cell. Models have been proposed for the aggregation of subunits of protein and polysaccharide via a "linkage" protein. Demonstration that the synthesis of acid mucopolysaccharides depends on RNA synthesis and is blocked by actinomycin D provides evidence that genetic activities are required for the process. The stability of proteinpolysaccharides in tissues has not been well established, but there is evidence that some mucopolysaccharides of the ground substance have a turnover rate of about seven to ten days. Lysosomes containing acid hydrolyses have been shown to be operational in many cell types in the processes of breakdown. Furthermore, studies in rapidly growing systems indicate that a delicate balance must be maintained between the synthesis and breakdown of various cellular elements. Degradative processes, however, may not always need the synthesis of new degradative enzymes. The activation of nonactive enzymes permits a rapid initiation of degradation. We consider that the observation that excess vitamin A increased the proteolytic activity of chondrocytes and released a protease from lysosome particles which was capable of degrading the protein fraction of the proteinpoly-saccharide, is of great importance from a phenomenological point of view, since similar processes may be operative in tumorization of the cell. The delay of this process by hydro-cortisone was finally interpreted by the authors as stabiliza-

tion in intracellular lysosomes. We should like to point out
here that Weber et al. have shown that the endocrine system
plays a major role in the control of glucose metabolism. For
example, glucocorticoid acts as an inducer, and insulin acts
as a repressor for glucogenic enzymes [52]. Insulin also acts
as an inducer for several enzymes and plays a co-ordinating
role in metabolic reactions as indicated [54]. The evidence
that glucocorticoid increases the synthesis of enzymes involved
in glycolysis suggests that this corticoid might produce a
stabilizing effect by the increased synthesis of extracellular
substances, as well as the simultaneous stabilization of
lysosomes.

6. SPECULATIVE CONCLUSIONS

On the basis of model-system analysis and the presented
review material, which is diverse and sometimes confusing,
we shall attempt to carry out some speculative analyses on
possible operational mechanisms leading to cellular transformation. It appears that one should be cautious in comparing
cells growing in a tissue culture with cells growing in the
tissues of a functional organism. Alteration of environmental
conditions can lead to rapid adaptive changes in the cells, and,
consequently, we are dealing in a tissue culture with a modified
cell in an artificial environment. The longer the cells grow
under these conditions, the more permanent and drastic become
the changes in their growth characteristics. It does not seem
surprising, therefore, that cellular transformation gradually
takes place, and that cellular metabolic and synthetic patterns
are altered. It is clear that cellular growth in tissues depends
not only on intracellular synthesis, but also on the local tissue
environment, which is also under remote control by distant
endocrine secretions, liver proteins, migratory cells, etc. The
organism as a whole possesses an organized system of
interactions, and failure at the level of a local organ or a
tissue can have drastic effects on cellular growth in different
parts of the organism. This is especially so when changes occur
in tissues which provide regulatory compounds via circulation.

Therefore, conditioning of tissue cells in one region of the organism can be affected by cellular functions in other areas. For example, one can say that while local tissue cells are "semiautonomous," they are subject to conditioning from distant sources. For example, specific liver proteins, such as gluco-, lipo-, and mucoproteins, may play important roles in conditioning various tissue cells in the system. Of primary importance in cell transformation, on the level at which it establishes itself as an independent cell, are the changes in surface structures which lead to cell dissociation. Chemically or virally induced cancer cells exhibit deletions or additions of surface components, a reflection perhaps of the genetic activity pattern. Since migratory macrophages and leucocytes exhibit characteristics similar to those of malignant fibroblasts, it would be important for detailed studies to be carried out on the metabolic and synthetic events taking place during the differentiation and developmental processes of migratory cells.

It appears that the synthesis required for the formation of adhesive intercellular linking is not operational, nor is it limited in the cancer cell. Furthermore, absolute concentration levels of the polymeric inducers in the intracellular space may be important, since these may indeed program synthetic processes in a normal cell aggregate. It seems that close proximity between adjacent cells is essential for the maintenance of concentration for small inductive molecules. It seems possible that the lack of maintenance of external microenvironment, which is under internal metabolic control, may lead to the initiation of degradative processes and cellular dissociation. For example, hypothetical repressor compound R_0 (Fig. 140) can be effective in suppressing synthesis and division only when cells are closely associated. It is our impression that muco-, lipo-, and glucoproteins are instrumental for cellular adhesion and organization into complex cellular structures. The function of these viscous molecules seems to be, in the first phase, to act as physicochemical complexing agents but subsequently they are instrumental in maintaining the genetic activity pattern. Since polysaccharides and starches exhibit a chemotactic effect, the monomeric carbohydrate subunits may have a leading role in inductive processes. Furthermore, in a normally growing cell the system of the proteolytic enzymes may be repressed and perhaps stored in the inactive state. Only when the cells move apart may derepression occur, leading to the formation or activation of various proteolytic enzymes. This, in turn, could lead to further cellular dissociation. It is well known that cancer cells in tissue culture exhibit highly proteolytic activity and, moreover, the degree of tumorization can be measured to some degree by this proteolytic activity [53].

While much nutritional information has been obtained in tissue culture studies, it is not possible to correlate this information directly with data obtained from cellular growth in tissues which are a part of the organism. Tissue cells which have been separated by proteolytic digestion have lost an essential part of their extracellular structure. The initiation of new growth and cellular aggregation may require supplementation of certain growth factors, as well as adaptive processes by the cells themselves. The formation of protein-polysaccharide type cross-linked fibrous structures in the intercellular space requires a complex set of enzymes. It is uncertain whether parts of this cross-linked proteinpolysaccharide material are obtained from distant sources via circulation—from the liver, for example—or normally growing and developing tissues, synthesizing everything locally. Since many proteins are derived from the liver, it would not be surprising if some specialized cells utilized proteinpolysaccharides which reached intercellular spaces via the circulatory system. These polymers could be assembled or aggregated under the influence of local cellular processes into a cross-linked polymeric network. The fact that some cells require glucoproteins in tissue culture growth strengthens such a hypothesis, but one should note that isolated cells may have been rapidly altered and thus require supplemented media. Since these glucoproteins are present in plasma at a higher concentration than normal during neoplastic diseases, it would seem that if these glucoproteins were utilized, the local cellular enzymes and substrates, which are instrumental in cross-linking, etc., would not be operational in a cancer cell.

It would be of great importance to determine the origin of the intercellular binding materials. This information would help to elucidate the mechanisms leading to intercellular adhesive cross-linking and would provide insight into the differential behavior of normal and transformed cells. While reviewing various sources of information concerned with cellular cross-linking and adhesiveness, one gets the impression that a potential neoplastic cell has lost the capacity to organize the polymeric network of proteinpolysaccharides, even though the basic elements might be present. This may include the loss for specific inductive enzyme synthesis (perhaps via polymeric inducers), which may be essential for the assembly process leading to cross-linking and cellular attachment. It further appears that maintenance of cellular adhesions may indeed depend on the balance between the formative and degradative processes. Such a balance could be displaced by reduced synthesis or by excessive degradation, later resulting, perhaps, from processes leading to excessive activation of degradative enzymes. While proteolytic activity would cause the destruction of intercellular proteinpolysaccharide or lipoprotein networks, it is not clear how these enzymes would initiate the degradative processes. The evidence that many protein-degrading enzymes break down intracellular adhesive structures emphasizes the importance of the protein moiety in the polymeric network. How this polymeric proteinpolysaccharide or lipoprotein network would be linked on the specific attachment sites on the cell surface is not clear. Since sialic acid type receptors are present in at least some cell surfaces, one could speculate that inductive enzyme synthesis produces monomeric enzyme units, which combine into di- or polyvalent enzymes (also containing inhibitory sites), which could act as cross-linking units between elements of the cell surface and the fibrous network. Since external structures on the cell surface seem to be under internal metabolic and synthetic control, we can assume that disorganization of the internal metabolic system will have reflections on the cell surface. There might be changes in the complexing properties of cell surface elements or alterations in the enzymatic pattern.

How could such functional disorganization in cellular metabolism and synthesis lead to abnormal growth? Model-system analysis has revealed that when an external agent interferes with cellular elements, there is a multiple set of possibilities to produce disorganization in the functional system. This phenomenon may also explain why cancer can be produced by a large variety of agents and types of interactions. In abnormal growth the principal aspect of disorganization would be the interference with the regulatory mechanisms which control the concentration and distribution of metabolites, internally as well as externally. Consequently, the selective, organized genetic activity pattern, which was developed during successive stages of differentiation, can be lost. During the progressive disorganization, cellular survival many depend on many interlinked factors. If the cell survives, it may become stable again, except that it will have lost many regulatory features which are present in the highly differentiated cell. The evidence that acquired characteristics of the tumor cell can be reversed [13-15] shows that we are indeed dealing with an intracellular regulatory process. Experiments with the model-system on the computer have shown that a functional disorganization can be produced at various sites, but it can be reversed at suitable conditions. Therefore, the trigger mechanisms which operate and manipulate major constitutional elements, such as RNA, proteins, and pools, can produce major functional changes in the cell. Consequently, a progressive change may occur once the chain mechanism has been started. Computer experiments have also shown that if the system has strong regulatory features, the system has more stability and can sustain drastic external or internal interference. Further-more, such a system cannot be altered rapidly. Studies by Pitot and associates have shown that alteration of template stability is one of the features of the cancer cell. How can this property be related to a state of disorganization? We should like to generalize and say that functional stabilities of various elements depend on cooperative interactions. Instability reflects the reduction of some essential cofactors in a cooperative system. Events leading to a reduction of cooperative processes represent disorganization. In the case of cellular injury, we have a wide range of disorganizations. In a tissue, for example, we actually have a large spectrum of cells which can have different degrees of disorganization and different potentialities for becoming cancer cells, but only a *specific conditional disorganization pattern* may lead to the cancer cell. The potentiality of the cell being induced to the cancerous state may depend on the state of the cell, the age of the cell, the number of generations removed from the embryonic state, etc. It also depends on the environment of the tissues and the functional stability of the endocrine system. Furthermore, the age of the whole organism can affect the neoplastic potentiality. This influence results from the imbalance which is developing in the endocrine characteristics during the aging of the organism. We could then say that a cell which exists in the aging organism, in which the endocrine system is losing its regulatory balance, is more likely to be more sensitive to disorganizing effects than a cell in a young organism.

One could ask how an external agent, such as a hormone, could influence the cellular growth pattern leading to neoplastic transformation. It appears that hormones which affect the metabolism and synthesis of cross-linking polymeric materials can have a modifying effect on cellular interaction. For example, one could postulate that hormones which suppress carbohydrate metabolism and synthesis could potentially produce a reduction of polysaccharides which are essential for adhesive intercellular networks. Hormones which affect specific synthesis of proteins and lipids may be similarly instrumental. Thus, under prolonged influence of such hormones, cellular adhesiveness could be reduced, leading to cellular transformation as well as cellular migration. Also, any hormone or agent which would interfere with the action or synthesis of enzymes which participate in the formation or maintenance of adhesive structures could also be effective in processes leading to neoplastic transformation. One can make a general speculative statement that regulatory substances, such as hormones and their analogs, can be potentially the most effective cancer-producing agents, when one considers that they operate at extremely small concentrations. The selective action of hormones on specific tissues would make these tissues targets for the cellular transformation processes. An excess of activating or a lack of inhibitory hormones could lead to an imbalance of the synthetic and activation processes, which are essential intercellular cooperative functions. It appears that a potentially important research area in the cancer field is not at the level of single cells or isolated cell suspensions, but in studies on the mechanisms of synthesis as well as degradation in the community of cells (tissues and organs).

7. SPECULATORY CONSIDERATIONS ON INHIBITION OR REVERSAL OF ABNORMAL CELLULAR GROWTH

The problem of whether abnormal cellular growth can be suppressed or inhibited in a biological system without interfering with normal cellular growth is a difficult one. Is a differential and selective therapy possible? It is evident that in order to achieve differential growth suppression, one has to have a considerable amount of detailed information on metabolic mechanisms of both systems. At the present time, the basic information in regard to detailed functional processes in abnormal and normal cells is limited, but the principal difference in major areas seems to be quantitative and not qualitative in character. Furthermore, cellular transformation represents a multitude of states, and not necessarily two distinct growth systems. The purpose of this discussion is to formulate some broad speculative suggestions.

We do not consider here the interactions by agents which suppress normal and abnormal growth simultaneously, having only minor differences, perhaps, in rate processes. Therapy based on such mechanisms is not effective on a long-term basis. It appears that one of the objectives could be a selective, planned disorganization of regulatory and synthetic processes of the transformed cell. Cellular regulatory systems have operational characteristics which are adaptive, and, therefore, when some metabolic pathway is affected by outside interference, compensatory mechanisms may become operational and minimize the effect. This suggests that a multicomponent interference process may be more effective in creating metabolic conditions which are more difficult or impossible for the cell to compensate by regulatory adjustments. Obviously, detailed knowledge about the system is required in order to plan and carry out such a cellular growth interference operation. The object of research should be to elucidate detailed differences in metabolic pathways between normal and abnormal cells. For example, the difference in carbohydrate metabolism between normal and transformed cells offers, potentially, some possibilities for differential growth inhibition, provided that the difference is substantial.

A multiple-step interference, however, could be more promising. From the point of view of the single cell, differences in metabolic patterns in cancer and normal cells may be more quantitative than qualitative. There could be differences in membrane structures and transport characteristics. While alteration of membrane characteristics may permit the cancer cell to increase its growth potential, it may also permit entry of various selective growth inhibitors into the cell at a much higher concentration than into the normal cell. Consequently, "natural" inhibitors, which operate in the regulation of normal cell growth, could be introduced into the cellular system. An increased intracellular concentration of inhibitors or repressors could perhaps be selectively built up in cells which have "leaky" membranes, while, in normal cells, the permeability barrier prevents excessive inhibitor concentration. As a result, a selective growth reduction or regression could be produced differentially in cancer cells. In order to gain entry into the cell without grossly interfering with cellular functions of all cells, natural growth inhibitors or their analogs seem to have the highest potentialities. This suggests that specific tissues and cell lines may require specific inhibitors. It may also be possible, however, that all cells which originate from the same embryonic cell may still have common basic growth and division inhibitors, such as has been considered conceptually in the model-system, for example, as in Fig. 140. In addition, one may speculate that genetic systems in higher organisms, developed during their evolution, may still have some common rudimentary genetic characteristics with lower species, but these genes may never become operational during the selective differentiation processes. However, they could become operational when genetic activation occurs during disorganized transformation. Consequently, species at a low evolutionary stage may be a source of potential inhibitory compounds. Besides selective growth inhibition, one can consider that a cell could be selectively modified or that the transformation could be reversed. Since the cancer cell is usually located within the mass of a normal tissue or circulating in a system where normal cells circulate, it is obvious that the action on the cancer cell must be highly selective in order to avoid interference with the whole system. It has been shown by Braun [15] that the cancer cell in plants can reverse itself. when exposed to certain growth conditions in normal tissues, but it is obvious that this type of procedure is more important in research than it is a practical procedure in cancer therapy. The very fact that a cancer cell exists within the system containing normal cells indicates that, under these conditions, normal cells have lost their potentialities to influence the cancer cells effectively.

There appear to be several facets to cancerization. Alteration of surface adhesive characteristics, for example, which is probably associated with the enzyme system and gene activation pattern, may be of primary importance. Certain primitivization of nutritional requirements, indicative of the loss of certain differentiated features on the growth regulatory level, may also be important. While carbohydrate metabolism has been extensively explored in cancer cells, it may also be important to look into nitrogen metabolism. It is well known in micro-biology that when bacteria transfer from one type of media to another, there is an adaptive change in the enzyme pattern. The question is whether the cancer cell has nutritional characteristics sufficiently different from normal cells to make it possible to block metabolic pathways in areas which are not operational in normal cells. Perhaps in the future, when we shall have more information about growth regulatory processes in the cell, it will be possible to plan a metabolic program for inducing the selective reversal of a transformed cancer cell.

REFERENCES

1. V. R. Potter, Cancer Research, 21: 1331 (1961).
2. H. C. Pitot, Cancer Research, 23: 1474 (1963).
3. H. C. Pitot, Perspectives in Biology and Medicine, 8: 50 (1964).
4. H. C. Pitot and S. Ch. Yoon, Biochem. Biophys. Acta, 50: 197 (1961).
5. H. C. Pitot, Cold Spring Harbor Symposium on Quantitative Biology, Vol. XXVI, p. 371 (1961).
6. H. C. Pitot and Ch. Heidelberger, Cancer Research, 23: 1694 (1963).
7. H. C. Pitot and C. Peraino. J. Biol. Chem., 239: 1783 (1964).
8. S. Ch. Yoon, H. C. Pitot, and H. P. Morris, Cancer Research, 24: 52 (1964).
9. T. T. Puck, in Growth in Living Systems (New York: Basic Books, 1961), p. 181.
10. H. W. Fisher, T. T. Puck, and G. Sato, Proc. NAS, 44: 4 (1958).
11. I. Lieberman and P. Ove, J. Biol. Chem., 233:637 (1958).
12. H. Eagle, Science, 148: 42 (1965).
13. A. C. Braun, Harvey Lectures, 56; (New York: Academic Press 1961), p. 191.
14. A. C. Braun, Ann. Rev. Plant Physiol., 13:533 (1962).
15. A. C. Braun, in Growth in Living Systems (New York: Basic Books, 1961), p. 567.
16. J. H. Lippincott and B. B. Lippincott, Science, 147: 1578 (1965).
17. J. P. Greenstein, Biochemistry of Cancer (New York: Academic Press, 1954).
18. A. Tyler, in Analysis of Development, (Philadelphia: W. B. Saunders, 1955), p. 556.
19. J. Holtfreter and V. Hamburger, Ibid, p. 230.
20. D. P. Costello, Ibid, p. 213.
21. A. A. Moscona, in Growth in Living Systems, (New York: Basic Books, 1961), p. 197.
22. M. H. Moscona and A. A. Moscona, Science, 142: 1070 (1963).
23. M. Abercombie, Ibid, p. 188.
24. R. Dulbecco, Science, 142: 932 (1963).
25. G. Klein and E. Klein, Ann. N. Y. Acad. Sci., 63: 640 (1956).
26. L. Purdom, J. Ambrose, and G. Klein, Nature, 181: 1586 (1958).
27. G. Klein, in New Perspectives in Biology, p. 267, Elsevier Publ. Co.
28. A. S. G. Curtis, Exp. Cell. Res., Suppl. 8, 154: 107 (1961).
29. D. W. Fawcett, Ibid, p. 174.
30. C. Grobstein, Ibid, p. 234.
31. P. W. Brandt, in Symposium on the Plasma Membrane (New York Heart Association, 1961).

32. A. C. Taylor, (1961), Exp. Cell. Res., Suppl. 8, 154: 161 (1961).
33. H. Eagle, Proc. NAS, 46: 427 (1960).
34. L. Levintow and H. Eagle, Ann. Rev. Biochem., 30: 605 (1961).
35. I. Lieberman and P. Ove, J. Biol. Chem., 233: 637 (1958).
36. R. C. Hughes and R. W. Jeanloz, Biochem, 3: 1535 (1964).
37. J. H. Highberger, J. Gross, and F. O. Schmitt, Proc. NAS, 37: 286 (1951).
38. M. R. Pollack, in The Bacteria, Vol. IV (New York: Academic Press, 1962), p. 121.
39. P. Hoffman, A. Linker, V. Lippman, and K. Meyer, J. Biol. Chem., 235: 3066 (1954).
40. S. A. Barker, G. I. Pardoe, and M. Stacey, Nature, 204:938 (1964).
41. R. C. Hughes and R. W. Jeanloz, (1964), Biochem., 3, 1543.
42. K. Schmid, J. Am. Chem. Soc., 75: 60 (1953).
43. R. J. Winzler, in The Plasma Proteins, (New York: Academic Press, 1960), p. 309.
44. R. J. Winzler, in Chemistry and Biology of Mucopolysaccharides, (Boston: Little, Brown, 1958), p. 245.
45. A. Gottschalk, Ibid, p. 287.
46. R. Y. Stanier, J. Bact., 54: 339 (1947).
47. M. O. Suda and Y. Oda, Med. J. Osana Univ., 2: 21 (1950).
48. B. M. Gesner and C. R. Jenkin, J. Bacteriol., 81: 595 (1961).
49. J. F. Riley, The Mast Cells (Edinburgh: Livingstone, 1959).
50. J. F. Riley, Lancet, 40: 7425 (1962).
51. S. F. Jackson, in The Cell, Vol. VI, (New York: Academic Press, 1964), p. 387.
52. G. Weber, R. L. Singhal, and S. K. Srivastava, Proc. NAS, 53: 96 (1965).
53. G. Cameron, Tissue Culture Technique, (New York: Academic Press, 1950), p. 109.
54. G. Weber, N. B. Stamm, and E. A. Fisher, Science, 149: 65 (1965).
55. H. Harris, Exp. Cell. Res., Suppl. 8, 199 (1961).

Author Index

Subject Index